DEAR
DAD

DEAR DAD

KATHY WILLIAMS

Deeds Publishing | Athens GA

Copyright © 2016—Kathy Williams

ALL RIGHTS RESERVED—No part of this book may be reproduced in any form or by any electronic or mechanical means, including information storage and retrieval systems, without permission in writing from the authors, except by a reviewer who may quote brief passages in a review.

Published by Deeds Publishing in Athens, GA
www.deedspublishing.com

Printed in The United States of America

Library of Congress Cataloging-in-Publications Data is available upon request.

ISBN 978-1-944193-20-1
EISBN 978-1-944193-21-8

Books are available in quantity for promotional or premium use. For information, email info@deedspublishing.com.

First Edition, 2016

10 9 8 7 6 5 4 3 2 1

This book is dedicated to my Mother who always wanted me to pursue writing and to my Dad to whom I made a promise that is now fulfilled.

ACKNOWLEDGEMENTS

A very special thanks goes out to my professor from Georgia Southern, Dr. Darin Van Tassell, who encouraged, inspired, and gave me the time to compose this memoir. The ability to inspire is a gift that you give to all of your students. To my classmates in Research Methods, you gave me input and support that were invaluable and you have no idea what that meant to me, even though I was the old person writing letters to a dead guy. Dr. Thomas, from my WWII class also deserves credit for his interest and support.

I am so appreciative to Lance Coalson of Father and Son Publishing who gave me hope, support, and sound advice.

Thank you to all of those whose input was invaluable to the content of the book; Dr Danette Boyle and Gene Little from OMA, Brenna Williams from the 4th Calvary Spur, Dr. William Allison, US Army War College, Phil Goldfarb who researched Philip Good Buffalo, as well as Joe C. Daniel who researched military records.

Then there is the wonderful Blue Ridge Writer's Group that gave me John Edwards and in turn Alice Eachus. You just never know where life will lead you in a kind word or a brief thought that leads to something remarkable and yes, somewhat strange and weird. Thank you.

I feel so lucky to have Pam Krazenburg as my editor. Her insight and suggestions were invaluable. She brought a new perspective to the book and challenged me to think deeper about the entire process and what it meant to me. Thank you Pam.

I feel so fortunate to have made contact with Bob Babcock of Deeds Publishing. How strange life can be. Our connection is astonishing to say the least. My father obviously held Bob in high regard to have kept mementos from their relationship all these years. The number of photographs presented a challenge in the book design that Matt King and Mark Babcock aptly met. Thank you!

I am lucky enough to have a really great brother and sister who have probably suffered in silence while I prattled on about the book but who never let on. They joined me in joy and sorrow and were kind enough to accompany me to the OMA reunion. How I love you both.

Thank you, Martha Gasoline.

A LETTER HOME

Dec. 5
Germany

Dear Mother, Dad & kids,
    Just a short note to let you know I got your letter dated Nov. 15. I also got 3 letters from Patti and a package from the Ackers so my birthday was pretty good with all those letters and the package. Tell the Ackers I really appreciated the package and if you'll send their address I'll thank them myself. I guess the packages will start coming in now I hope.

— 2 —

I made myself a beautiful foxhole today. I dug a hole about 4 ft deep and seven ft. long and covered it with logs and dirt. I got a battery and rigged up an electric light — its really swell. Tonight I'm lying here writing, reading and eating while shells are falling all around but I just laugh at them from my foxhole deluxe.

That clipping you sent of Evelyn Davis was another Evelyn Davis or else she

—3—

has a new face.
 Please keep sending as many packages as you can as I'm sure they'll catch up and I'll really appreciate them.
 Yes it would be nice if the war were over by Xmas but seriously the way things look now I would not be surprised if we were still fighting next Xmas. Their whole army including the air force is very much stronger lately. In my opinion (and I ought to

—4—

know) much of their equipment is just as good if not better than ours and they sure know how to use it.

Well best wishes for a Merry Christmas and a Happy New Year, I'll be thinking of that big turkey and I hope next year I'll be with you.

Love to all,

Jack

PREFACE

In order to better help the reader navigate through our book, understand that I am picking up where my Dad and I left off when we began to read his WWII letters. Some of his letters are grammatically erred and I left those errors intact. As I read his letters I find that something will trigger a memory, a story or event that happened long after his time in WWII. Sometimes I feel compelled to tell him stories about my life he had no idea about. These are the things that, for some reason or another, were left unsaid. Then I write to him. I do not really understand why I feel the need to talk about some of these experiences. I find that I sign my letters with nicknames. His favorite one for me was "Martha Gasoline," a play on my real name, "Martha Kathleen." To share only my father's letters without giving some insight as to the man he became would be a failure on my part. Do not doubt that I carry a bias of love and admiration. It is my hope that somehow, as inept and inexperienced as I am, that you, the reader, will come away with a greater understanding and admiration for my father. It is also my hope that you will learn from my regret, the regret that I did not spend more time asking my father about his life instead of spending so much time absorbed in my own.

CONTENTS

Part I 1
February 8, 2014 3
January 1943 9

Part II 13
To England June 1944 15
July 1944 23
August 1944 29
September 1944 47
October 1944 69
November 1944 97
December 1944 113
January 1945 127
February 1945 161
March 1945 177
April 1945 199
May 1945 211

Part III 243
May 1945 after the War 245
June 1945 253
July 1945 267
August 1945 273
September 1945—Helsterbach, Germany 279
October 1945 285
November 1945 291
December 1945 297
March 1946 301
April 1946 307

Epilogue 311
Afterword 319

WHO IS WHO

Aunt Betty: Sister-in-law to Dad. She married Dave.
Dad: General Judson F. Miller
Dave: Dad's brother
Doug: Oldest son of Dad and Mother
Fred: Dad's youngest brother
Juddy: Youngest son of Dad and Mother
June: Dad's second wife-they had no children
Kathy: Oldest daughter of Dad and Mother, the one writing to Dad
Leroy: Kathy's second husband
Liberty: Oldest daughter of Kathy
Loralei: Youngest daughter of Kathy
Mimi: Mother to Dad, Dave and Fred. Married to Samps. Her real name was Martha Davidson until she became a Miller.
Mother: Married to Dad. She was an Army nurse. Her name was Bette Marguerite Lancaster.
Ross: Loralei's husband
Ryan: Liberty's husband
Samps: Father to Dad, Dave and Fred. Married to Mimi-his real name was Herbert F. Miller.
Shelley: Second daughter of Dad and Mother
Travis, Ross and Skylar: Children of Loralei and Ross. Kathy's grandchildren.
Travis: Son of Kathy

PART I

FEBRUARY 8, 2014

Dear Dad,

It is 8 pm; I am sitting here with our customary glass of wine staring at my cell phone still expecting to see your phone number light up the screen. It only seems rational to continue our tradition-you sitting on your porch in Washington and me sitting at home in Georgia unwinding from the day. I've chosen my best bottle of White Zinfandel; I know you preferred Chardonnay. Catching up with you was the only "happy hour" I looked forward to every day. It has been two years since your passing, two years of regret, remorse, and grief. But you would be proud of me today, Dad. I know you said you were always proud of me yet today I finally gathered enough courage to open the box you sent me years ago. I will admit it was not easy, it feels so final to be going through your letters. Thank you for trusting me with the treasures of your life. I know how important these letters were to you and I promise to write your book. Except now it will be our book, all our unfinished conversations

Love, Your daughter Martha Gasoline

February 8, 2014

Dear Dad,
 Do you believe in reincarnation? I know that is a weird question to ask a person who is dead but there is something weird and kind of creepy happening with my cat. She is obsessed with your letters and the box where the letters are stored. I know you are not a big fan of cats and to know that you have been reincarnated into a cat would be an awful twist of fate for you. But I have a 16 year-old cat named Gracie; you met her on your last visit to Georgia. She had never taken an interest in your box of World War II letters until just

recently when I officially opened the box and started reading them. For a couple of weeks, I had some of the material on my coffee table and the damn cat started sleeping on it. She never used to go on the coffee table. I was worried that she might have a "Gracie cat rage fit" and start shredding things (which she has done to magazines and newspapers in the past for some unknown slight that I surely must have committed against her). In angst I organized some of the material, placing it back in the box and pushed it under the computer desk. Now the damn cat is infatuated with the box, constantly sleeping and grooming on it.

She has quit going on the coffee table. I don't suppose you have come back reincarnated in my cat, have you? We do have a "CAT" history.

Do you remember when I was younger, perhaps 8 or 9 years old, we lived in Germany and I had two kittens? You tried to carry them off but they returned, undeterred by your efforts to dispose of them. You were also steadfast and stubborn, I might add, and repeated the abduction of my kittens to parts unknown. Your plot was once again foiled as the intrepid twosome navigated successfully back to their kitten loving girl. You gave up. Then In Ft Meade, Maryland, I got into trouble when I stowed 2 different kittens in the basement storage of the apartment. You did not scold me but this time the kittens really had to go, or we would have been evicted.

If you have returned to this world of the living as a cat, you came to the right place. You get milk every morning, Fancy Feast every night (Elegant Medleys as a matter of fact), and brushed twice a day. In the middle of the night when you open and close the cabinet door in the bathroom, I know that is my command to let you out. You even prefer and get your water from the shower floor.

Love, Your girl who brings home kittens

February 14, 2014

Dear Dad,

Why is it that some people know their purpose in life at such a young age while others take a lifetime to find happiness and fulfillment? You always wanted be an officer in the military. As I dug into the box I came across some really interesting pictures. Were you struck in the womb with this ambition? I love the Wild West look, so serious and deadly. You look like a good ole Sooner. Yet the picture that really foreshadows your future is the military pose with your younger brother Dave at your side. Even in childhood you appear disciplined, controlled, and confident, standing proud in your uniform. The pictures are cadet photos from Oklahoma Military Academy the "West Point of the Southwest" back in the 1940's. You were well on your way to achieving your goals there, fours years of high school and then two years of junior college at this prestige military academy. But then the US entered World War II and the urge to reach the front lines and fight for your country was too compelling.

As soon as you turned 18 you enlisted as a buck private, the lowest rank, and left school during your freshman year of college. Poor Mimi and Samps, this decision was not what your parents had in mind. While it was extremely brave, it was not a good career move. The plan was to graduate from OMA as an officer. Instead you would be going to war and would not be safe.

The first step was Officer Candidate School at Ft Riley, Kansas. One of your fellow candidates was a guy named Oleg Cassini. What was a famous designer doing at Ft. Riley Cavalry School at that time? You said Oleg struggled some but was a really great guy who had a good looking girl friend. However you were most impressed with the girlfriend's sister, "a real babe" according to you! Was Oleg Cassini's girlfriend the famous actress Gene Tierney?

Another sign was your favorite colors were red, white, and blue. How lucky you were to have had such a presence of mind and intent so early in life. Your granddaughter, Liberty, also discovered her passion as a child. She seemed to know at the age of thirteen that she wanted to be a veterinarian. Despite a lot of girls who say they want to be veterinarians, my daughter Liberty pursued her dream with a determination and sense of purpose that I believe must have been inherited from you.

Love, Your girl and proud mother

Wild west look

With younger brother Dave

OMA cadet photos

JANUARY 1943

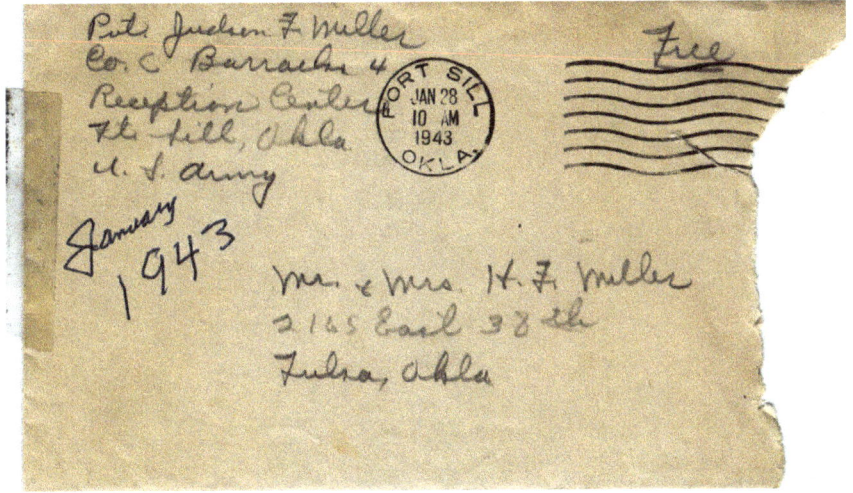

Wed. night January 1943

Dear Mother & Dad,

    I meant to write you yesterday but I haven't had a lot of time. I'm in the same barracks with the other O.M.A. boys. Junk and I are the only ones who have our uniforms so far. I got a pretty good fit. I'll finish my shots and tests tomorrow. I sure hope I get shipped out soon. All we do down here is sit around except when we are being processed. I saw Pat O'Hornet today, he's a Sgt. I imagine I'll be shipped out by Monday at least. The average time for a man to stay here is 5 days. I had a fine trip down with the Sweeneys. We didn't get here till about 9:30 that night. If I can't make Officer's School something will be wrong. I never saw so many dopes in all my life. They are terrible. The food down here is swell. I like everything fine but I'll sure be glad to get out of Ft. Sill. Please write often as I'll be looking forward to hearing from you. My address is:

Pvt. Judson F. Miller
Co. C Barracks 4
Reception Center
Ft. Sill, Okla.
U.S. Army
Be sure you address it right. I'll write as often as I can.

Love, Jud

CLASS ROSTER
37th OCS

GRINAGE, MARION K..........................*President*

Agardy, Louis
Agnew, William R.
Austin, Daniel D.
Austin, Donald R.
Barnes, Robert V.
Beeman, Gerald W.
Blend, George W.
Blodgett, Robert A.
Borg, Robert J.
Brown, Jacobs A.
Campbell, Bruce M.
Cassini, Oleg L.
Centrello, Angelo
Childs, Charles E.
Clucas, Edward W., Jr.
Coffey, Ulysses M.
Cromwell, Raymond F.
Cunningham, Leo P.
Deines, Julius
Diemer, Roland M.
Downer, John T.
Embry, Richard L.
English, Edward L.
Fleur, John H.
Gallert, Earl W.
Gay, David S.
Girton, Leroy M.
Gustafson, Erland W.
Hall, Fred M.
Halpern, Samuel A.
Harlan, James A.

Harris, Robert A.
Hazeldahl, Loren B.
Hughes, Howard M.
Hutchinson, Warner C., Jr.
Irving, Clark O.
Jackson, John P.
Johnston, Charles R.
Kelly, Frank B.
Koester, Fred
Krueger, Louis T.
Langwick, Charles
Mahan, Thomas L.
Mezga, Joseph C.
Miller, Judson F.
O'Connor, Joseph E.
Penley, Gerald H.
Price, Ernest V.
Ramsey, Eugene P.
Schirding, John H.
Sharp, William D.
Smith, George H.
Suman, John G.
Thrush, Francis H.
Wampler, John J., Jr.
Watson, Doma A.
Watson, Roy E.
Welch, John M.
White, Lawrence J.
Wilkins, Fred
Willson, Harold, Jr.

THE CAVALRY SCHOOL
FORT RILEY, KANSAS

/ks
12 Aug 1943

In Reply
Refer to: 201 – Miller, Judson F.

SUBJECT: TEMPORARY APPOINTMENT.

TO: Judson F. Miller,
2d Lieutenant, AUS

A   01032728

1. The Secretary of War has directed me to inform you that the President has appointed and commissioned you a temporary Second Lieutenant, Army of the United States, effective this date, in the grade shown in the address above. Your serial number is shown after A above.

2. This commission to continue in force during the pleasure of the President of the United States for the time being, and for the duration of the present emergency and six months thereafter unless sooner terminated.

3. There is inclosed herewith a form for oath of office which you are requested to execute and return. The execution and return of the required oath of office constitute an acceptance of your appointment. No other evidence of acceptance is required. This letter should be retained by you as evidence of your appointment.

BY COMMAND OF BRIGADIER GENERAL RAMEY:

D. W. JONES,
Lt. Col., AGD,
Adjutant General.

INCLOSURE:
Form for oath of office.

Ft. Riley, Kans., 2-18-43—1682—3000.

Accepted appointment 12 Aug 1943.

PART II

TO ENGLAND JUNE 1944

To England June 1944

V-Mail 1

Dear Mother & Dad,

There isn't much to write. All I can tell you is that I'm somewhere in England. That's all we can say. I've seen quite a few interesting sights since I left Okla. But I guess I'll have to wait till I get back before I can tell you. Everything is fine so don't worry. How are the kids? Tell them hello. The English beer is quite a bit different than 3.2 in Tulsa. I'm having a heck of a time getting used to this money, too. My mail hasn't caught up with me yet. I haven't had a letter since I left home. Write soon.

Love, Jud

V-Mail 2

2nd Lt. Judson F. Miller
210th Rpl. Co.
c/o Postmaster, N.Y., N.Y

Dear Mother & Dad,

Nothing at all to write but I just thought I'd drop you a line. I got my first letter from you the other day since I left Meade. After this use V-mail please. There is absolutely nothing to do here except go to one of the pubs and drink beer. I don't like their beer near as well as I do ours. By the way whenever possible please send me some food, I'm about to starve to death. The other day I bought an English bicycle, it sure helps out. It's just like having a car in the States. I'm finally getting used to this money. For a while all I could do was to hold out a handful of money and tell the bartender to take what he wanted. Please write soon and be sure to use my new address at the top.

Love, Jud

V-Mail 3 June 24, 1944

Dear Mother & Dad,

    I just got your first V mail letter so I thought I'd drop you a line. I think I can tell you a little bit about my crossing. I came over on a former luxury liner, I can't tell you it's name but you've heard of it. I had a stateroom with some other officers—it was pretty nice. The food was the best I've had in the army. Best of all I didn't get seasick and it was plenty rough. Last night a couple of us rode our bikes over to see a famous cathedral, pretty interesting. By the way, I forgot to send your signature to the bank so I'll cable the money to you for the car. Glad to hear you've got it back. I'll continue this on another page.

Love, Jud

{PAGE 2}

Dear Mother & Dad,

    I can't say that I like this country. Everything is so old and out of date. The weather changes every minute, real cold or real hot. It doesn't get dark till after midnight and everything closes at ten. But it isn't too bad and I guess there are a lot worse places. I'm going to a cricket match this afternoon. I'll probably get a big kick out of it. One good thing about this place is that you can't spend much money—nothing to do. Tonight I'm going to the cinema as they call it. Sorry to hear about Fred's knee, must be working too hard. By the way Bill Potts is here, I see him every day. He is a Capt. Tell the kids hello and write soon. Potts and I are going to play tennis tomorrow.

Love, Jud

February 16, 2014

Dear Dad,

I called Aunt Betty yesterday. She is the only one left from your generation in our family. I told her about the book project and I asked if she could help me answer some questions. I also told her about the cat's obsession with your WWII papers. She told me that she had always heard that a cat will "seek out the person in the room who does not like them and then sit in their lap." Brouhaha! Who knows?!

The first V-mail you wrote details your crossing of the vast Atlantic Ocean and noted that the ship used for transport was a "former luxury liner" with a familiar name. Aunt Betty thought the ocean liner you were alluding to was the Queen Mary but she wasn't positive. I did some digging and discovered the Queen Mary was responsible for channeling more than 750,000 soldiers during World War II. I can picture you aboard that mighty ship which is now anchored peacefully in Long Beach, California.

Were you feeling homesick in your first V-mails sent from England? How could you not feel distress, surrounded by war and trying to adapt to a new country. Yet in your letters you are more concerned about the car. What was so important about the car? I remember the afternoons Doug and I spent reading some of your letters with you and the car was mentioned often. When I asked you about it, you said that if Mimi did not get those car payments she would have to get rid of the car

Aunt Betty filled in some of the missing pieces for me. Were you referring to the car you shared with Uncle Dave? He was allowed to use the car early in the evening while you were given the later evening privileges, being the oldest. Aunt Betty said you had a car accident causing damage to one side of the car. In an attempt to conceal evidence of a collision from Samps, you decided to park the car on the street with the damaged side hidden from view. Ha, ha good try!

You aren't the only one in our family to wreck a car. Remember when I was 16 and crashed our beloved Chevy Impala convertible? I know how much you loved that car. White exterior, red interior, and quite the bomb with the top down! Definitely a "Chick Magnet Car." We all thought we were so cool cruising around in our convertible. But the car did have a dangerous blind spot when it pulled up to a road stop that was not perfectly perpendicular. I sure was surprised when I hit that Cadillac. I called the house to tell you. Immediately you asked me with concern in your voice if I was all right. Then the tone of our conversation changed. Now I understand why you parked your wrecked car with the good side facing your house. It is an awful feeling knowing you have disappointed your parents. But we both took full financial responsibility. There was just enough damage accrued to our car and the Cadillac for the cost to be under our deductible. I spent the rest of the summer mowing lawns, babysitting, and house cleaning to come up with the $130.00 for the repairs. It was such a relief to hear you often repeat that when I decided to have a wreck I did it right, with a Cadillac. Ha!

Love, Your Cadillac wreck girl

V-Mail June 27, 1944

Dear David,

    I just got your letter and sure was glad to hear from you. Glad to hear that the car is fixed. This place is sure different from the States. It rains all the time and it's always cold. I live in an old home built in about 1800. What a dump, no hot water or anything and I have to do my own laundry. There isn't a darn thing to do here, the one and only movie changes once a week. All the pubs close at 10. I sure will be glad to get out of here. Got to close now. Thanks for the letter and please write again.

Your brother, Jud

February 18, 2014

Dear Dad,

Your brother Dave says the car is fixed. Hmm, was it a mechanical issue or the infamous damage that you tried to hide from Samps? The mystery deepens. I am trying to feel sympathy for you when you are forced to do your own laundry. But I can't help but remember your sarcasm when we whined about our laundry. Poor baby (that was what you always told us). When I read this to my sister Shelley she also said "Poor baby." What happened to "Get tough or die?" Ha. Ha. Who did your laundry at OMA and OCS? Did you guys have maids or something? No hot water and the movies only change once a week!! So much for the comforts of home. Maybe we take too many things for granted sometimes.

                Love, Your girl who always did her own laundry

P.S. I am sorry all the pubs closed at 10.

Dad's first convertible

JULY 1944

V-Mail July 4, 1944

Dear Mother & Dad,

Well, here it is the 4th of July, sure doesn't seem like it. Its just like any other day here. I'm getting so I dislike England and the English more every day. The country is about 50 years behind the times and the people are convinced that they are practically winning the war alone. I've got an easy job here now. Until I ship out I'm mess officer here at this camp. All I do is sit in the mess hall all day. There isn't much to eat though—by the way, how about that box you were going to send. Also please send some single edge razor blades, these over here aren't any good. I sent a money order for the $25 for this month yesterday, you should get it soon. Today the sun came out and it stopped raining for the first time in 9 days. Please let me hear from you soon. Almost forgot, please subscribe to the World & Trib. and have it sent to me.

Love, Jud

February 19, 2014

Dear Dad,

You were a mess officer? Ha! I am trying to picture you in a chef hat or apron or something but it doesn't work. Chef Judson. You do become a griller supreme. I can remember when we traveled from Ft Leavenworth to Oklahoma to visit Mimi prior to moving to Germany, we would stop and grill lunch on the side of the road. I had no idea you had prior experience in culinary arts. There were many scenic places that had shade and picnic tables. At one stop we even found a turtle that you allowed me to take to Mimi's. I was in charge of the large picnic basket. You taught me how to use Sterno to fire up the portable stove so you could grill the hamburgers. Fast food was not the thing that it is today. No FFDF (Fat F****r Diet Food) menus available.

Your crowning culinary achievement has to be your homemade spaghetti sauce. No one, no restaurant has ever made a sauce to compare. Did you formulate this sauce while waiting for the war? You know your spaghetti sauce has become a family tradition. Even as adults visiting, spaghetti was expected to be on the family menu the first night. Shelley and I have continued the tradition. Every year when we run the Gasparilla Distance Races Shelley makes "The Spaghetti." I probably shouldn't tell you this, but she has adulterated one vital ingredient of late. She now makes it with tofu and ground turkey instead of ground beef. I realize that this is heresy for you. You and Mimi were blood-thirsty carnivores who ate raw hamburger meat. Does this satisfy some deep primitive primal urge? As for me I want all meat burnt. I can still see you reach into the refrigerator and come out with a wad of raw hamburger meat saying "Yum, Yum" and laughing at my contorted face as you gorged on the nasty wad. Yuck. The first time I saw Mimi do this I was equally appalled and horrified, especially as I was still quite young and impressionable. I probably have some psychological scaring or something as a result.

Love, Your girl still traumatized by raw meat

July 27, 1944

Dear Mother & Dad,

Sorry I haven't written but I've just been too busy. I am now somewhere in France. I've really been in some interesting places lately. I wish I could tell you about them but I guess it will have to wait till I get home. By the way use the address on the envelope. These French are really interesting to talk to if one can understand. I'm picking up a little—my Spanish helps a lot. I am now living out in the country in my pup tent. We bathe in a creek near here—no hot water. We eat field rations—same thing every meal every day—it really gets old. I just got back from a farmhouse near here. We got several bottles of apple cider. They would rather have cigarettes than money. I haven't spent a franc since I've been here. We get our cigarettes and soap free—plenty of both. I now have about 6 cartons of cigarettes that I use to trade for cider and congac. It rains all the time—everything I own is wet but I don't mind it so much. It's twice as rough over here as it ever was at Gruber. The thing that gets me is that there is absolutely nothing to do in our spare time. In England at least we had movies and pubs. All towns are off limits over here. We don't have lights or anything. My letters may be a little further apart now but don't worry. How about a box of cookies, candy, and stuff. I'm afraid I'm going to lose a lot of weight on these K rations. I got my head shaved right down to the scalp—I really look good. You ought to see my tent. Five of us put our tents together—we've all been together since we came over. We have our foxholes right inside the tents. Please write often.

Love to all, Jud

February 20, 2014

Dear Dad,

    Now it started to get real. Still no hot water and lived in a pup tent with a foxhole as your friend. I wonder if you missed that old 1800's house in England. However it appeared your "hunter-gatherer" skills were productive. Apple cider and cognac!! Well-done, Dad!! Who needs the pubs to stay open after 10 pm? K rations do sound yucky. So you asked your mother to send cookies, candy, and stuff. You do realize that these are FFDF. Today that is a huge problem; all these people are on the FFDF diet.

    You shaved your head! This is the style nowadays. Both Doug and

Travis have shaved heads and they don't have to worry about lice (Ha, Ha), which may be why you shaved your head since you were coping without the niceties of life: hot water and regular bathing. It is supposed to be very macho. Actually Travis started shaving his head because he was cheap and did not want to spend his limited college student funds on haircuts.

It is interesting to read that your Spanish helped you communicate with the French. I am still working on my minor in Spanish. I will finish this December, finalmente.

        Love, Your girl who refrains from FFDF

AUGUST 1944

August 4, 1944

Dear Mother & Dad & Kids,
    I just ran out of ink so I'll have to use pencil. I am now right up at the front in a regular outfit. I'm in a tank troop, which suits me fine. Our casualties are a lot lighter than most. By the way, this is German stationary. I'm enclosing a few German papers, maybe dad can read

them. I've got a luger pistol and a lot of other stuff. The things I hate worst are the air raids. They usually come over at night just when you are going to sleep. Then everyone starts shooting at them and wakes you up. We either have to dig a foxhole and sleep in it or sleep in our tank. You ought to see my tank. I've got bottles of French cider or congac hanging on the turret and the people throw flowers all over our tank. It really looks gay. Everything we own is strapped on somewhere. About money—I haven't been paid this month yet but as soon as I get it I'll send it all home—we can't spend it here. If the war lasts long enough I'll come home rich. You ought to see all the equipment "Le Boche" leaves behind. Tanks, guns, and everything. The other day we went riding on German horses and then on German motorcycles. The food is too bad, the other day we—shall we say found a chicken and fried it—really tasted good. Today is a beautiful day, one of the best yet. We are back of the lines waiting to move up and kind of taking it easy. Please don't worry if you don't get a letter every week. I'll try to write but sometimes it's impossible. By the way I have a new address.

Troop F, 24th Reconnaissance Squadron
A.P.O. 230
c/o Postmaster, New York

Please write often, your letters are really appreciated. Tell the kids I'll bring back a pistol or something for them.

Love to all, Jud

August 5, 1944

We're waiting to move up today and just laying around. Please get the subscriptions to the papers started, and also send over such magazines as Life etc. I could sure use them on a day like this. We just came through a pretty good-sized town and we were the first Americans to go through. They almost smothered us with flowers and tried to drown us with cider & cognac. How about some food also. One thing—if you have a camera around that you could spare and some film I could sure take some wonderful pictures to bring back. So please see what you can do.

Love, Jud

March 2, 2014

Dear Dad,

I can only imagine what you were feeling. I know that you were required to censor the letters of your men and you also had to be careful of what you wrote. I also believe that you were trying to protect your family from the horrors you were experiencing. In that respect perhaps censorship becomes a means to free you from any responsibility you may have felt in revealing the entire truthful brutality of war. How can any of us really understand your reality without being there with you? It is just not possible. I feel like my life has been so insignificant in comparison. I would think that the waiting would be the worst, leaving your imagination free to go places that may not be so wonderful. I am constantly amazed how you conveyed a feeling of nonchalance to your parents and brothers. You always seemed to find the bright side of it all; getting rich because you can't spend any money, your cache of cognac and cider, not to mention your hunting and gathering of "le Boche" leftovers. I would give anything to have a picture of you and your tank draped in flowers, cider, and cognac!

You rode German horses?? Gee, nothing like going to Europe to go horseback riding! I know that at OMA horses were a big part of your training. Is this where my "horse craziness" came from? I found a picture of you on a pony. What a cutie you were!

You took me on my first real horseback ride at Ft Knox for my 6th birthday. You rode a chestnut horse that was very barn sour, but no match for you. My horse was a black and white pinto; I can still see us going down the trail. I was hooked and announced I was going to be a jockey when I grew up.

When I was riding at Blue Ribbon Downs in Sallisaw, Oklahoma I rode for a guy named Jack Steele, among others. We usually rode at a brush track called Mel Mar Downs. Jack Steele was a very interesting looking man with a Freddie Kruger like scar that started from his ear, went under his chin, across his throat, to his other ear. Yet his scar was not the result of a horror movie but the enemy trying to cut his throat during the Korean War. His unit was overrun during the night as they slept and nearly all were killed. The story goes that Jack feigned his death and survived. You were also in Korea. I remember you telling me about your time in Korea, lamenting it was the only time you had ever known the US Army to go hungry. You told me about how some of the men, (you did not participate) would take the frozen corpses of North Korean soldiers and arrange them around your camp so that if you were overrun during the night they would be shot first, allowing more time to fend off the attackers. I guess Jack's bunch didn't have any frozen dead guys to stick around their camp. Sounds macabre but useful.

Did you ever meet a guy named Jack Steele? He was supposed to have been married to a woman who was the heiress to the Smirnoff fortune. Who knows? Could just be a silly horse track rumor.

So you "found a chicken" did you? Hmm. Playing hide and seek or what?

Love, Your wanna be jockey girl

August 7, 1944

Dear Mother & Dad,
Not much to write but we're kind of taking it easy today so I

thought I'd write. The other day we came upon an abandoned German supply dump. When they left they didn't take a thing with them. They left clothing, boots, rifles, machine guns, grenades, and a million other things, including some delicious canned fish of which we got a whole case for our tank. I'm sending home some bayonets and things as soon as I get a chance. I'm enclosing some photos I found there, too. French civilians were picking up clothes by the wagonload. One man picked up a grenade, pulled the pin, and just tossed it aside. Several people were pretty badly hurt. I had just left before it blew up. We even captured a German automobile—we had a big time driving it around and running into trees.

What I wouldn't give to sleep in a good bed again. A tank is so close to the ground so that when you sleep under it you can't turn over. The darn planes keep you awake most of the time. I haven't had a bath since I've been in France. Boy a hot shower would really feel good. Nothing else to write so I'd better close. Please write soon.

Love, Jud

March 3, 2014

Dead Dad,

You will never guess what kind of typing error I just had. I typed Dead Dad not Dear Dad. What does this signify? I am well aware that you are dead. The brain is a strange mechanism. For example I have been taking Spanish for the last three years and it bleeds into my English so that sometimes a bunch of garbage comes out of my mouth. The other day in my Research Methods class I was trying to say "documentary." But guess what pops out instead? Some stupid sounding mix between "documentario" and "documentary" that was incomprehensible. The class must think I am an idiot. I know I felt like one. My spelling is taking a hit, too. I find myself writing a Spanish word when I should have used the English word and vice versa.

It pains me to read that you were smelly and nasty from not bathing, eating the spoils of war, which in your case is canned fish, and then joyriding German cars into trees. It must have been awful to see the grenade explosion and the aftermath of human life. But what were you thinking driving the German car into trees? Is this reckless disregard for human life hereditary? I think you should know, years after the fact, I found out that my brother Doug, your son, let my thirteen-year old daughter, Loralei, drive him from Tampa to Valdosta on I-75. That is a 4-hour drive! She did not have a license and as far as I know had never driven a car before except on the farm. I wonder what else I have yet to find out.

Love, Loralei's mother who knows nothing

V-Mail August 12, 1944

Dear Mother & Dad,

Nothing much I can write but we're kind of taking it easy right now so I thought I'd drop you a line to let you know I am okay. I'm getting pretty sick of this fighting but it could be a lot worse. The only time the "Luftwafe" bombs us is at night and our bombers are after them all day. You ought to see all the dead animals laying around. They get all bloated up and smell awful. Most of the towns are just a heap of rubbish but the French are glad to see us. When we take a town they throw flowers and give us cider, congac, calvados, & wine. Their calvados is 135 proof. I had my first bath in a dirty lake. I don't know which was dirtier, the water or me. I am waiting for some magazines, newspapers, and the other stuff I asked for. Don't worry.

Love, Jud

March 8, 2014

Dear Dad,

I guess the French must have had a lot of alcoholic beverages. I seem to remember from my French classes in high school that the water in France was supposed to be bad and that is why everyone drank wine, even at a young age. I am a little suspicious of that explanation but I have heard worse justifications for indulging in inebriating activities. I had never heard of Calvados until I read your letter so I was amazed that man's ingenuity extended to making booze out of apples. Maybe that was the real reason for God forbidding Adam and Eve from eating apples. Who knows? Now the current fad is to make apple-flavored beer.

I am trying to imagine the dead and bloated animals laying everywhere and it seems grotesque, Picasso's Guernica comes to mind. You did not say anything about dead people. I think you were shielding your family but I wonder how you really dealt with this.

       Love, Your girl who prefers to eat apples not drink them

P.S. I have a confession to make. When I was around the age of 17 we (my erstwhile Bay Crest friends in liquor cabinet crime) used to take small sample portions out of each kind of booze that our parents had in their liquor cabinets. We would combine these in to a big bottle and then go to someone's house when the parents were gone. We poured this noxious concoction into a large punch bowl that we called the "Suicide Bowl." Then we commenced to get stupid drunk. This was followed by stupid sick.

August 15, 1944

Dear Mother, Dad & kids,
Excuse this torn up German stationary but it's all I can get. We're

kind of resting up today so I have a little spare time. I sure wish I had a camera—I could really get some interesting pictures. These French people are a scream—they kiss you and every one of them feels like he has to shake your hand—I feel like a politician trying to get votes. Back in England the kids all said, "Any gum, chum" over here they say, "Gotta franc, Yank" or "Cigarecte por papa?" or "Bon bon" which means any candy. For a package of cigarettes you can get two-dozen eggs and we get plenty of smokes for nothing. Consequently I'm getting awfully tired of eggs. The weather recently has been hot in the daytime and cold at night. The other day we stopped by a river and got a bath—it sure felt good to be clean again. Last night we came through a town and a Frenchman (Frogs we call them) gave me a bottle of delicious red wine which I am drinking now. I've got a bunch of souvenirs that I'll send as soon as I can get a box, including a German canteen which I'll fill with cognac. We even use the darn stuff in our cigarette lighters—no fooling.

August 18. I had to stop in the middle of my letter and this is the first chance I've had to finish it. I found out about my pay today—since I didn't get paid last month I'll draw 2 months pay the last of this month. I'll be sending home something like $250 and you can use that on the car. How much more do I owe on it? I'll sure be glad to get it out of the way.

I saw in an army newspaper where General Lear says we'll probably go to the Pacific after we finish here—that sounds like a rotten deal to me. My mail still hasn't caught up with me. I'm looking forward to some letters and the other stuff I asked for—newspapers, magazines, food, and a camera & film if you can find one.

Right now I'm cooking some hard boiled eggs in our little gasoline stove we carry in our tank. About half my spare time I spend censoring the mail of my platoon. Please write often and don't worry. I'm just fine except a little hungry and I sure would like to sleep in a bed with a roof over me.

Love to all, Jud

March 12, 2014

Dear Dad,

So you have lots of cigarettes and eggs at your disposal. Is this how COPD and high cholesterol begin? I do not remember you dealing with high cholesterol but you did develop COPD towards the end. You really hated it when they put you on oxygen and you had to drag that long oxygen tube all over the house with you. The day you went into the hospital for the last time you refused to wear oxygen and I ended up carrying the tank myself. You walked in, the essence of the soldier, your back straight, step firm, and bearing crisp and commanding. No one who witnessed you stride into the hospital that day could have imagined that you would never walk out.

As I remember it, you preferred your eggs all runny and gooey. Yuck. As for me, I have to have mine cooked all the way through. We had a maid in Ft. Leavenworth when Juddy, my youngest brother, was born. She was very sweet, an elderly black lady who told me that baby Jesus would be happy if I would please eat my eggs, even if they were runny. I did not buy her argument and even though I felt bad about refusing her requests I just could not eat the "baby Jesus runny eggs." Even today when I see runny eggs I think of "baby Jesus runny eggs." Mother used to try the argument, "the children in Europe are starving" guilt to get us to eat stuff we did not like.

Two days before you went into the hospital for the last time you and I went to Johnny's on the Dock. We each ordered the crab omelet. It was wonderful, not only the food but to be there with you. I need to go now Dad. Sorry.

Love, Your girl who needs a good cry

March 13, 2014

Dear Dad,

    Sometimes when I try to write to you I do just fine and then sometimes like the other night I get the "Forever Sadness." Grief is such an exhausting, wrenching emotion. It catches you at the most unexpected times. One day not too long ago Shelley was going through some of Mother's things while cleaning and organizing. She had a "Forever Sadness spell" and called me, "Do we ever get over losing these people?" I told her no, we never do but it is far better to have our precious, exquisite grief than to feel nothing. There is something godlike about grief, something so holy and pure that at times I even welcome its arrival. To lose Mother and then you 77 days later was so hollowing, so emptying, that I felt like I no longer existed. I am so very sorry that I could not bring myself to call you right away when she died. I did not want to speak about it, not to anyone, not even myself. But when Shelley told me you cried when she told you, I felt like I had wronged you in some way by not telling you as soon as it happened. I forgave you many years ago about what happened between you and Mother; this had nothing to do with my inability to talk about her death. I cried most of that night. The next day I knew I would go crazy if I stayed home so I went to work like some kind of a zombie workaholic person. I think you would have understood that. Work has always been my refuge from dealing with anything emotional. I go and hide in work where things make some kind of rational sense. Well, not always but you know what I mean.

                                            Love, Your girl with the Forever Sadness

March 15, 2014

Dear Dad,

    Today I ran the Gate River Run in Jacksonville. It is a 15K. For me, the best part of the race is around the 9th mile, where you trudge over a huge bridge known as the Green Monster that spans the St John's River and ends at the football stadium. Every year they have music blasting from loud speakers on the Green Monster in an attempt to alleviate our suffering and take our minds off the incline. This year there were military running chants, as well as the usual rock and roll music. As soon as I heard them I thought of you. Doug had bought a tape of running chants for you. Every year when we visited it became tradition to play the chants on our way to Paradise on Mt Rainier. My favorite one was:

"Two old maids lying in bed
One rolled over and the other one said
I want to be an Airborne Ranger
Live a life of guts and danger
Airborne, Airborne
Airborne, Airborne, Airborne!"

Sometimes when I am at races and I get beside some military units I will start chanting this one. They will often chime right in with me. How great is that?!

I see in your letter that you enjoyed a glass of red wine, I wish I had a picture of that. Even more interesting is the blatant smuggling of cognac via a German canteen to your parents. Ha, ha! It is really distressing albeit amazing, however, that you used cognac for cigarette lighter fuel. I hope you did not later regret this when you needed a nightcap and had no more booze. I wonder if Mimi had started to use her dishwasher as her liquor cabinet by then or was that a later development? I remember the first time I saw this unique dishwasher utilization. I was quite impressed that my grandmother was so ingenious. Booze in the dishwasher and raw hamburger meat in the fridge! What a gal!

Love, Your girl who uses a dishwasher for dishes

"Mooriew"
Ashwell Lane
Glaston bury
Somerset
20-8-44

Dear Mrs. Miller,

I thought you would be pleased to hear my friend & I met Jim & Jud one evening. They both came in for coffee. We discovered they were both living next door to us. Jud was anxious for you to know exactly where he was & asked me to write to you but as I have been very busy I am afraid my letter was delayed.

Jim came round one evening whilst I was in the midst of icing a Wedding Cake & he was very interested & helped me with the decorating. He was telling me how he had made preparations to be married but was shipped off a little too soon. Myself I thought he was far too boyish to be married.

We had arranged for a picnic one evening but the boys had a slight problem so we couldn't make it. They have been gone about 3 weeks & I hope by the time this letter reaches you war with France will be over.

The weather has been marvelous during the last two weeks. This place seems dead now with no boys around. There are about 5 girls to every boy at the dances so we have really decided not to go unless something different happens around here. With best wishes to you & hoping the boys mail is reaching you & that everything is alright with them.

Yours Sincerely,
Marion Parfitt

March 21, 2014

Dear Dad,
This letter from the English lady, Marion Parfitt, was very nice. Now I know that you were in Glastonbury Somerset before you shipped. This letter must have meant a great deal to Mimi because she kept it all this time. Mimi was really something

I loved her deep voice even if she did sound like a man, she was so petite and tiny and then she would speak. It was very disconcerting to some people the first time they heard her. She would always laugh about how people reacted, telling us that on the telephone the person on the other line always referred to her as sir. She looked too sweet to be someone who eats raw hamburger.

She always wore high heels. She warned me to be careful about wearing high heels too much. It seems that she had worn them so long that late in life she was unable to walk flatfooted. She even had high-heeled slippers. She demonstrated this to me and ever since then I have had a thing about avoiding high heel shoes.

Her collection of cigarette holders would have made Marlena Dietrich envious. They were in all lengths and colors. I always thought she was very glamorous and sophisticated. Her wit was the best and she would always find the humor in every situation. When it came to a lively political conversation, she could hold her own with the best of them. Every time I would visit she would tell me about the time I told Samps to go stand in the corner during a fancy cocktail party. I must have been about 2 or 3. One of the distinguished guests asked me why Samps had to stand in the corner. I told the guests that it was because he had wet his pants.

One of Mimi's best stories was about the time that she and her sister, Lil, were on a trip to Venice, Italy. They were taking a gondola ride in the Grand Canal when Lil suddenly violently sneezed, ejecting her only set of dentures into the canal, never to be seen or heard from again! The rest of the trip Lil had to tough it out, sans teeth.

Another sister of Mimi's, Aunt Margie, used to put spiked eggnog in her mashed potatoes. That's a pretty ingenious way to booze it up. You were named after Mimi's brother, Luther Judson Davidson. He was the captain of the SS Rochester, the first ship sunk by a German submarine off the coast of New Jersey during the war. Luckily he survived. Mimi referred to her other brother as, "that SOB Carlisle." I never did find out why he was "that SOB." Before I understood that "SOB" was a derogatory term I just thought he had a weird name. After all, the rest of Mimi's siblings were named Minor, Rose, and Semple.

Did Mimi ever tell you about the time that she and Samps had an argument and she uttered the curse word "damn" and threw the cigarette ashtray at the couch? I bet that scared him worse than being put in the corner for wetting his pants. She could be very intimidating even at the height of maybe 4'11"?

It is now cocktail time, according to Mimi, so I am going to pour a glass of wine and toast the two of you.

        Love, Your girl terrified of high heel shoes

Dad on a pony

Jack Steel and Wanna Be Kathy

Mimi and Dad

Dad & Samps with namesake Uncle Jud on ship deck

SEPTEMBER 1944

Somewhere in Belgium—September 6, 1944

Dear Mother and Dad & kids,

It's been quite a while since I've written but as you've probably been reading in the papers we've been traveling pretty fast and I just haven't had any time. So don't worry whenever you don't get a letter. I won't tell you much about my experiences because if I told you everything I'd be writing for a week. I'll just save them and tell you when I get home.

The other day we shot up a column of German trucks, two of them were full of perfume and face powder the Germans had stolen from Paris. I got a whole bag full of assorted stuff. Last night I washed my feet in "Evening in Paris" perfume. I'm going to send some to you and Patti if they will ever let us start sending packages again.

The weather is getting awfully cold and hardly a day passes that it doesn't rain. Everyone is really miserable. I've got an awful cold and it looks like I'll keep it till the war is over because you don't ever get a chance to get warm or dry. All my blankets and clothes are wet and full of mud—you can't even build a fire to dry them out. Sometimes I think we ought to do like the Germans do—move into the peoples' houses.

You ought to see these German prisoners we take. They are all scared to death because they think we are going to shoot them, which wouldn't be a bad idea. We caught an officer and 3 privates the other day. The officer refused to sit in the same jeep with his men. We had to knock him around a bit and then he was very happy to sit down anywhere. The SS troops are the worst ones. They are all a bunch of fanatics. They shoot at you till they are out of ammunition then come out yelling "Kamerad"—we usually shoot them. But when you take one prisoner, they start crying and are scared to death.

I haven't got paid yet but it should be in any day and I'll send it home the minute I get it. You ought to see the way the people treat the collaborators. They shave all the women's hair off and shoot the men. The Germans sure treated the people badly and they are really glad to see us. I've only gotten mail once since I've been in France and I sure would like to get some more. Write often and don't worry.

Love, Jud

March 29, 2014

Dear Dad,

August 15 to September 6 must have seemed like an eternity to both you and Mimi. According to a book of yours that I kept, *Hitler's Last Gamble,* the General you were serving under was known as "Lightning Joe Collins." There was little time for writing. I also kept the framed map that was always on the wall from my earliest childhood on and, until now, I never really understood its significance. What a treasure it is for me now!

I have to say that you truly continue to amaze me. You captured German trucks loaded with "Evening in Paris" perfume along with face powders. Ah, the spoils of war!!! The luxury of bathing one's feet in expensive Parisian perfume!! It does make sense to bathe your feet in something that could be helpful in preventing trench foot since you write about how wet everything is, even if it seems a little sissified. You are no sissy.

It was interesting to learn how the German officers seem to have some sort of snobbery in regards to fraternizing with their men. I later thought more about your reaction to the snobbery and arrogance of the German officers. I remembered my first lesson in humility that you gave me. We were living in Bad Kissengen, Germany. A girl, named Kathy Acres, rudely threw an apple core at me, hitting me in the face saying, "Apple core Baltimore, who's your friend, who's your enemy." I attacked with all that a scrawny, asthmatic girl of 9 could do, which in this instance was pretty inadequate. She pulled a wad of hair out of my scalp and I began to bleed profusely. She beat the crap out of me. Certain of my imminent death, or at least a bald spot for the rest of my natural life, I ran screaming home to my Mother. I was extremely indignant not only because of my wounded pride but also that I had suffered this humiliation at the hands of a Sergeant's daughter, while I was the daughter of the commanding officer, a Major. How dare she attack me! What a little rank snob I was. You were kind but firm in your lesson. I was so ashamed. You had the Sergeant come to the house with Kathy Acres. We had to apologize and shake hands. My bald spot was only temporary. My lesson in humility, arrogance, and the temptation to abuse a position of power has stayed with me to this day.

I found some pictures of German prisoners in your papers. I am not sure of the timeframe but I guess that is not really that important. The look in their faces says it all. They look so young.

Too bad fanatics are still around today. Must be one of those forever things. They just morph depending upon the culture. I sometimes wonder if humans are really worth saving. Then something noble and honorable happens and my faith in the good is restored.

Love, Your girl who did nothing significant at age 19.

Somewhere in Belgium—September 15, 1944

Dear Mother & Dad & kids,
There is plenty to write about but I just don't have the time. Today we are working on our tanks and cleaning weapons. Once in a while they pull us back for a day or two to take care of equipment.

I don't guess it will hurt to say that the part of Belgium I'm in now is awfully close to Germany and most of the people are Germans and also Nazis. It sure is a lot different from France where the people were all so glad to see you. Here they stay inside and don't seem very happy to see you and already some of our men have been shot by civilians. Right now I'm in a house that belongs to some Nazi general who left for the Fatherland when we came. It really is fancy—he even left a big roadster in the garage—we've been driving all over in it. Tonight if we are still here I'm going to sleep in one of the beds for the first time since I left England.

I've been seeing a lot of action lately and killing plenty of Krauts. The soldiers we've been fighting recently are these S.S. troops. They are plenty good till the going gets rough and then they start yelling "Kamerad" and crying. A few days back we caught a bunch of them in a town and cut them off. We killed 23 and captured about 15— Boy there were dead Krauts all over the place. I was standing up in the turret shooting them with a machine gun and pistol. I should have been down inside—one of them shot at me and grazed my hand—nothing to worry about, it hardly drew blood. It sure made me mad though, I blew his head off with a burst from my machine gun. One I shot with my pistol sure was hard to kill. I shot him four times with my pistol and thought he was dead when he raised up and yelled "Kamerad" so we backed up the tank and ran over him. I don't take many prisoners because they are just a pain in the neck till the MPs take them off your hands. If you could see what the S.S. does to civilians you wouldn't take prisoners either. One Belgium family of 8 was locked in their house by the SS and then they threw grenades inside and set the house on fire.

So much for the gruesome part of the letter. I got paid and as soon as I can get a money order I'll send it home. I haven't gotten any packages yet and I'm really looking forward to them. The weather is awfully cold and a few nights ago we had a terrific cloudburst and haven't seen the sun since. That's all the time I have. By the way, how about the newspapers and magazines?

Love, Jud

April 4, 2014

Dear Dad,

I hardly know what to say. I have never had anyone try to kill me. I never blew anyone's head off and while I have briefly contemplated running over some idiot with my car in a moment of road rage, it has never become a reality, not yet anyhow. I try to envision squishing a guy with your tank. The closest thing that comes to mind is squishing those humongous cockroaches that go crunch when you stomp them. I have hit wildlife on the road unintentionally, and felt terrible about it. It really ruins my day when that happens.

I do try to stop and pick turtles up on the road to get them off the highway. Once, while in Oklahoma, I saw a huge turtle and stopped to get it off the road. He or she was the biggest turtle I had ever seen outside of a zoo. As I stooped down to pick it up it suddenly lunged at me with its cavernous mouth wide open, scaring the crap out of me. Astonished, I realized that it was a snapping turtle bent upon my destruction. I was still determined to save this savage creature, so I found a large stick. I gingerly poked it in the direction of the beast's gaping maw. "Snapping Turtle Creature" grabbed the stick and locked its jaws like a pit bull. I then successfully dragged the ungrateful beast to the side of the road and cursed it better than any sailor could have done.

Love, Your girl who saved the life
of an ungrateful Snapping Turtle Creature

P.S. I hope you got to sleep in the bed that night.

Somewhere in Germany—September 17, 1944

Dear David,

I got your letter yesterday and was very glad to hear from you. I'm glad you and Fred had a good time back East.

The Germans seem to be fighting a lot harder now than they did in France or Belgium. I nearly had my tank knocked out the other day by a 60 ton Panther tank, he hit all round us but fortunately he missed. He is too big for my tanks to fight and the only thing we can do when we meet one of them is to get out of the way. The gun on his tank is as long as our whole tank. It sure is a helpless feeling to sit there and shoot and hit him and he just keeps rolling towards you. What is more fun than anything is to catch a bunch of Krauts in a town. We move in and really tear the place up, shooting high explosive shells in windows and getting the Krauts with machine guns when they run out to surrender. I especially enjoy finding German soldiers in a German town because it gives me an excuse to burn the place to the ground. At night all the German civilians have to stay inside or else get shot—quite a few have already been killed that way. Whenever we kill a bunch of Germans we carry on our favorite sport we call "ghouling"-taking their money, watches, pistols etc. I have a slick cal.38 pistol I'm going to bring home. Its fun killing Krauts as long as they don't have any tanks—then it starts being fun for them.

Tell mother I got letter no.11 yesterday and that I have sent a money order for $230 home and I'll send more as soon as we get paid every month. I haven't received any packages yet but they say that they usually take quite a while so I'll probably get them soon. Please have the folks start a subscription to both papers as soon as possible. By the way, how is the car running? A couple of days ago we captured a Mercedes-Benz sports roadster and drove it all over till one of the men ran it into a tree.

The guys we are fighting now are the S.S. troops—they are the real Nazis and fight harder than any. I don't take any of them prisoners except once I went into a house after some and I ran out of ammunition for my both my tommy gun & pistol and I had thrown my last grenade when 8 of them came out of the cellar yelling "Kamerad"—I don't know who was the most scared—they didn't know I was out of ammo so I had to take them back. Well this is about all the time I have so I'll have to stop. Write soon.

Love, Jud

P.S. Something funny happened the other night. We didn't camp till after dark and we couldn't see what was in the next field. In the morning we woke up and found a bunch of Krauts camped not 20 yards from us. I'm glad we woke up early, we killed about 20 of them but the rest got away.

April 5, 2014

Dear Dad,

I am aghast. Ghouling? Yuck, but understandable given the circumstances. I do remember you conveyed admiration and respect for German tank engineering when we talked about your tank in comparison with what you were confronted with. It was formidable.

I remember us talking about this incident with the SS troops in the cellar. Good bluff, Dad. That was very lucky on your part. Not so lucky for those guys camped right next to you that didn't get up early enough. You were always an early riser.

I have never had to face what you did but I have been challenged many times. I was pretty scared the time I tried to summit Mt Rainier. When you attempt to summit Mt Rainier they put an avalanche beacon on you. Suddenly you realize that there is some serious stuff going on here. I pretty much pooped out at Ingraham Flats. I had become so fatigued that I felt I was endangering my fellow climbers. It would have been irresponsible for me to continue. There was also something going on with my left foot. As we headed back to Camp Muir and descended Cathedral Rock this 6 foot 200 pound guy falls on me, and we start to slide. We were all still roped together at this point. I did the group arrest thing, and yelled, "falling" which is just "Mountain Code" for "I'm gonna die." They teach you to say this so everyone on your rope will dig in with their ice axes and stop the slide. No one died; the big guy had blown out a knee. We all got off the mountain.

When I participated in my first triathlon I found another source of terror. It was held on Jekyll Island, the same place where we vacationed as we traveled from the Army War College in Carlisle, Pennsylvania to your new post at MacDill Air Force Base. You always took us to the most interesting places. I can still remember seeing an armadillo for the first time; I thought it was an anteater. Ha! You and I both laughed at my error. You always had a way about you that was devoid of ridicule or arrogance when you corrected or taught us, never belittling or humiliating us.

I had trained religiously and practiced the course twice. However, the ocean during my training had been tranquil. The day of the triathlon the waves were more like something spewing out of a vengeful sea demon's dark void. There was a tropical depression forming, and the officials were discussing whether to replace the swim with another run, not uncommon if the water is deemed too unsafe. However, it went off as planned, even though those in my group of newbies were obviously terror stricken. I know I was. What was I thinking? Idiot, you are 61 years old, go home and learn to knit or something. What if a shark eats you, idiot woman? But no, I go off with the rest of them. We were like lemmings going over a cliff; must have been herd mentality or something. I kept telling myself to calm down and just swim. I vowed to never do it again so long as I survived. As I rounded the last buoy to head for home I saw people doing CPR on someone in the rescue boat. Unfortunately he didn't make it. They pulled 6 swimmers out of the water that day. I did not need a rescue but I looked pretty awful; drooling sand out of my mouth and my eyes full of salt. You always told us to finish what we started.

Love, Your girl who knows how to scream "falling."

P.S. Travis did summit Mt Rainier the summer after you died. I know you must be proud of him. Sometimes I think that this was his way of dealing with your death. Do you remember singing "Candy's dandy but liquor's quicker" to him when he was the tender age of 16? I was appalled. Nothing like the granddad corrupting the grandson.

Germany — September 19, 1944

Dear Mother, Dad, & Kids,

Enclosed are several money orders totaling $230. Please apply this to my debts and it should pretty well square me up till next payday—God knows when that will be. But I figure that if I keep sending most of my pay home I'll soon have the car paid for and start saving a little of my own. I kept $75 of my pay this month in case I ever run into a quartermaster to get some warm clothing. They all stay so far back I doubt if they are off the beach yet. I'd sure like to get some clothes, they say it gets as low as 20 degrees below over here—that will just about finish me.

Please don't worry about me over here, these Krauts are very poor shots and besides, anything smaller than a machine gun just bounces off a tank. I'm glad I'm not in the infantry; they don't have several inches of armor plate to protect them. The only danger is from anti-tank guns and enemy tanks and when I see one of those I take off and clear out. But anybody who says he doesn't get scared when he goes into combat is either lying or else a damn fool. I've been in so many actions and battles I've lost count and I still get scared when we go into the middle of it. What gripes me is when I read about the "Lovely rest camps of the coast where combat troops go to relax"—that's a good one, we haven't had a day of rest since I've been in the outfit. The only ones who go there are the quartermaster and transportation corps, and the poor air corps who never have anyone to shoot at them and who sleep on cots and eat "A" rations, and go back to the States every other month. They do a darn good job but they sure don't deserve the extra pay and all the glory they get.

I hope you are right about the war being over in November but it doesn't look like it over here. The going is getting rougher & rougher. We are always reading where the Germans are disorganized and fleeing madly. Well I've run in to quite a few who never read our newspapers. And also the German people are not so happy to see us as the newspapers say.

I sure appreciate the clippings you send. Whenever we do have a little spare time there is absolutely nothing to do. We can't even go into a German bar because some soldiers were given some poisoned beer. So please get the newspapers and magazines started. Also I could use some cigarettes and single edged razor blades and food of any size, shape, or color. I guess I told you about the cow who didn't know the password and we had to shoot it—too bad—we had steak for 3 days after that. That's all the time for now, I may have time to write again this afternoon but I doubt it.

Love to all, Jud

April 6, 2014

Dear Dad,

Wow you are still paying for the car! This is such an ordinary thing to concern yourself with especially since you were in an extraordinary time and place. Perhaps that is how we stay sane when awful things are going on. We focus on the mundane, the ordinary. You seemed a little cranky in this letter. I don't blame you one bit. You didn't pretend to be unafraid or some kind of super hero; just honest and forthright.

I have been thinking a lot about being afraid lately, as your letters are sobering. There is intensity to them that was lacking before. I remember my worst fear, my youngest brother Juddy and the tragedy that would define our family. Fear comes in all sizes and shapes doesn't it? Let's move on for now and focus on something mundane or at least not so depressing.

Hmm. A cow did not know the password. We actually talked about this incident. You said he violated the perimeter, and did not know the password. Ha Ha!! You also told Doug and me that after the war was over, you were called to company headquarters. Your superior questioned you about a claim that a farmer had lodged over a cow that was shot by your unit. It seems his claim was valid since you put it in writing. I am amazed that anyone kept up with all of that. You also said that one of the guys in your unit was a butcher by trade. Now that was quite fortuitous!

Love, Your girl who cannot remember all her passwords

Germany—September 30, 1944

Dear Mother, Dad & kids,

Well, I finally got another chance to write. Please don't worry when you don't hear from me. I write every chance I get but the cavalry is always up in front and we don't have much spare time. Right now we are resting and cleaning equipment for a day or so, all we have to do is run a patrol every once in a while. I'm spending this day or so in unaccustomed luxury. We captured a town without much trouble so the electricity & water supply are okay. We took over the best house in town and I have a beautiful room with bath (hot & cold water), radio, soft bed, and everything. It probably won't last long but from now on I'm going to kick the Germans out and let them be uncomfortable instead of me. The weather over here is getting very cold and damp. I understand from the natives that it gets plenty cold in the winter. Right now I'm a gentleman of leisure. I have on some silk pajamas and bathrobe that belongs to the Kraut who lived here. I just got my monthly whiskey ration which officers get, it consists of 1 qt. of White horse Scotch and a qt of gin. I've got a big fire going and the radio is tuned to a U.S. dance band. Outside shells are landing not very far away, I'm scared to death the Krauts will hit the water main or power line—it would be just like them. Boy I'd like to finish the war right here, what a life—but I guess it will end in the morning or the next day, then back to my foxhole in the rain.

By the way, you might send me some stationary. The only paper I get is what I find on dead Germans.

One of my best friends (a lt.) was badly wounded the other day and another kid killed when a bazooka knocked out their tank. I was supposed to take my platoon on that patrol but they had changed it at the last minute—what luck.

I heard Churchill's message the other night. He said the British were doing as much fighting as the Yanks. Those dammed Limeys are about the sorriest fighters I have ever seen. Everyone here hates the Limeys as bad as the Krauts.

No I didn't go through Paris although we were mighty close to it. I have been in some of the famous campaigns but I can't say where yet.

I got a letter from Mrs. Love the other day asking about Wayne. Please call her and tell her I'll answer as soon as possible (I have a lot of loafing to catch up on while possible). I ran into a Lt. I knew from Wayne's outfit and he said Wayne wasn't hurt very badly. In a way he is lucky—no more fighting for a while anyway.

I haven't received any packages yet but they always take a long time.

I hate fighting in these big forests over here. The other day I almost lost my tank when I was going down a narrow road. I had infantry working with me and they found 6 mines about five yards in front of my tank. Don't worry I'm taking good care of myself, these Krauts start running fast when they see tanks unless they have tanks or AT (anti-tank) guns and then I'm plenty careful. Our equipment is better than the Krauts and one big advantage is we have plenty of ammunition. When we get in battle we fire as fast as we can and spray the country with lead while the Krauts *usually* have to take it easy.

Say about that West Point appointment, I sure would like to get one and if I do I can come home and get out of the army to go. So if you can do anything about it please do because I think 20 is the age limit.

Well that's all for now. I'll write tomorrow if possible. Please write often—letters mean a lot over here, and by the way if you see Patti tell her to start writing more often because I'm not writing any more till she starts writing more. If she is too busy running around with 4-F's to write I'm sure too busy killing Krauts to write.

Love to all, Jud

P.S. Every time we see an English soldier the Yanks always say, "Hey bud have you captured Caen yet" "Duffy's Tavern is on now" Another thing that tickles me is I'm too young to vote but I have to witness the men's ballots.

Here is some French & German invasion money we are paid with
1 mark=10 cents
1 Franc=2 cents

April 7, 2014

Dear Dad,

You were no longer a man washing his feet in Evening in Paris but upgraded to a gentleman of leisure in silk pajamas! Boy, you were almost considered a sissy! No, I take that back. You squished guys who refused to die with your tank and shot poor cows who could not remember passwords. I hope you enjoyed your jammies, hot water, and monthly White Horse Scotch ration. You deserved it!

It is somewhat sobering to realize that some of these letters of yours that I am now handling were written on paper that came off dead guys.

I am sorry to read about your best friend who was badly hurt. Did you feel guilty when the leadership changed which unit went on that patrol? Probably some responsibility mixed with relief then more guilt because you felt relief.

You wrote about fighting in big forests. The map we have points to your location in Hurtgen Forrest. I once saw a documentary that described how the trees became projectiles as they were blasted by enemy fire. I am glad you were protected in a tank.

Hmm, sounds like girlfriend Patti was ratted out by someone. Maybe Dave? I wonder if she wore Evening in Paris Perfume on dates with those 4-F'ers? Did her 4-F'ers know that you were off squishing Krauts with your tank?

I understand the insult about "Have you captured Caen" yet but the reference to Duffy's Tavern escapes me. Was it a soap opera or what? Let me know.

Love, Your girl who does not own silk jammies

P.S. I googled Duffy's Tavern and found out it was a silly radio show. It always began with a phone call to the tavern. The person answering would always say, "Duffy's Tavern, where the elite meet to eat."

Germany—September 30, 1944

Dear David & Fred,

"Goot Morgan" as they say in Germany. How is everything back home? It gets worse every day over here, the Krauts don't run away in Germany like they did in France & Belgium. So David you wish you were over here—I'll gladly trade with you any day. It isn't so exciting after you've been in combat a while. Its just like going hunting every day, you get tired of that after a while. The only difference is that the Krauts are hunting you, too. I've killed more Krauts than I ever killed rabbits. It's just a slaughter when we catch a bunch of them without any big stuff. In one town we killed over a hundred in the main street—there were guts, brains and blood all over the main street—we didn't lose a man.

I wish I could send all my souvenirs home—helmets, German tommy guns (Burp guns we call them because they fire so fast it sounds like someone burping), Lugers, rifles, and a whole pile of stuff I've had to throw away because I didn't have room to carry it.

I got a hot bath and put on my long underwear this morning so I'm all set for the rest of the winter. That's all the time for now. Write soon.

Love, Jud

April 11, 2014

Dear Dad,
There is not a lot I can say in response to this letter. I think that killing had become a necessary drudgery for you. It was also dehumanizing you. I have been thinking about this. I look at what I have been doing to make a living for the last 21 years and one thing that I always feel dirty about is the fact that I work for an industry that actively markets and sells tobacco products. The convenience store industry is bemoaning the shrinking of this highly profitable category while I have silently cheered it. I smoked for 8 or 9 years off and on before I finally beat the habit. Let's face it, the tobacco industry is the only business that I can think of that sickens and kills its customers. They are "Merchants of Death." It may not be killing in the same sense as what you are doing, but for me the killing that you are doing

is more honest than the killing that is done by the tobacco industry. There is no honor in killing for profit.

Love, Your girl who hates the tobacco companies

Map of Dad's journey

Surrendering

So young

Lovely mud

Kathy drooling sand

Travis thumbs up at summit of Mt. Ranier

OCTOBER 1944

October 1, 1944

Dear Mother, Dad & kids,

Well another month gone by and Uncle Sam owes me another $180. I probably won't get paid for a week or two but when I do I'll send it home. Pretty soon I ought to have the car paid for and start laying some away. The only good thing about being over here is that I don't spend any money. Please let me know about how I stand.

The weather is really bad over here. Honesty it has rained every day for the last three weeks. I sure would like to see it dry up. Its getting plenty cold now, too. The Red Cross sure does a good job over here. These Red Cross girls bring their clubmobiles right up to where the shells are falling and give us coffee and doughnuts. That's further up than the Quartermaster goes. The other day one of the men in the troop ran into his wife up here, he sure was surprised.

If you want to send any more packages here are some of the things I'd like to get. Cigarettes, candy, canned soup, magazines, cookies, stationary, more film if you can get it (I'm looking forward to the camera, and anything to eat—sardines would go good. That's about all for now. Write soon.

Love, Jud

April 12, 2014

Dear Dad,

The constant theme of paying on the car prevails. I remember that you and Mimi both talked about how she required you to start paying her room and board when you got your first job pumping gas at a local gas station. I think you were about 12 or 13. She must have been pretty tough as far as teaching you how to be responsible and disciplined with your money.

It strikes me as ironic that your first job in a gas station is going to be the same business that I will retire from in October of this year after 22 years of service. I never had any plans to work in this industry, it just happened at a turning point in my life. My marriage was in trouble and I had three children that I desperately wanted to see through college. Life as a racehorse trainer required me to move the family twice a year and was not very profitable. I needed stability for the children. Mother was actually the one who told me that a nice lady named Margaret was looking for help at the Food Flash store. It was really Flash Foods but Mother always got it backwards. What is even more ironic is that Samps was very much in the oil business. He started as a stenographer moving up the ladder until he retired 48 years later as a vice president and director of The Carter Oil Company. I found some copies of articles about him that I have included. The first article is from *The Link* June 1945 while he was still working. The others are about his retirement. I also found some really great photos. I can see some resemblance to you, although you kept your hair.

It is interesting to me to see a picture of you on his desk in his study. How proud he was of you! I thought it was funny to read that he owed his success in some part to having to leave the fruit business because of "an attack of jaundice attributable to excessive sampling of our products." Ha Ha!

There are many similarities in your story as well as mine to that of Samps. He started at the bottom. You started at the bottom, a private who became a General. I started at Flash Foods as a midnight cashier and worked my way up to District Manager.

Love, Your strong-willed girl

Samps

Samps looks like a choirboy

GENERAL JUDSON F. MILLER & KATHY WILLIAMS

Dad looks like the dandy Lord Fauntleroy

DEAR DAD

"I'd Like To Start All Over Again - - -"

By H. F. MILLER

H. F. Miller, manager of crude oil purchasing department for the Carter and a director of the company, has been a part of the Standard Oil Company (N.J.) organization for 42 years and 6 months. His friendly, likeable manner and ready smile have won for him wide popularity throughout the company.

SOMEONE asked me recently, "Now that you are approaching sixty, what have you gotten and what will you get for the years you have put in with your company?" Here is my answer:

I have thus far enjoyed more than forty-two years of association with a group comprising all sorts of personalities, but wholly motivated by a common purpose—the success of the company and our own individual well being.

My experiences and observations with regard to opportunity for promotion have been that this has always existed to a degree comparable with the company's growth and success, and that individual preparation for the assumption of increased responsibilities is never wasted. Milton's "They also serve who only stand and wait" can furnish consolation only to butlers, footmen and others with obscure futures, as it doesn't apply in the oil business.

I have enjoyed and am enjoying the mental comfort which comes with the knowledge that company-sponsored death benefits far more liberal than those enjoyed generally by workers in our own and other industries, are available to my family should I die before my retirement.

I have found that the difficult problem of adjusting salaries so that these shall be, all things considered, commensurate with the work performed and conditions prevailing, has been and is, for the most part, the subject of most careful study and consideration by those charged with this responsibility.

I find that at my age I am a capitalist in embryo and that when I retire I shall enjoy, so long as I may live, the fruits of my years of participation in the mutual efforts of the group comprising our shareholders, my fellow workers and myself, in the form of a monthly income from funds set up under company-sponsored plans.

I have found our company management in all phases of its endeavors to be constantly on the alert for and welcoming suggestions and expressions of opinion in regard to the company's functionings, but not seeking advice, which, after all, is the cheapest of all vices.

LIKED FRUIT

I like to reminisce and compare my present lot with what it might have been had I remained with my only other employer, a broker of dried fruits and canned goods in New York City, whom I served at the age of sixteen for a period of three months, my training for a business career at that time being limited to Horatio Alger's concepts of what a boy must do to succeed.

The deaths of the partners in this enterprise have long since brought about the dissolution of this small business. An attack of jaundice attributable to excessive

(*Continued on page 15*)

55% OF CARTER EMPLOYEES HAVE TEN OR MORE YEARS OF SERVICE

On the Carter payroll are 2,061 employees, of whom 1,129 have more than ten years of service—and this figure does not include the 365 persons now on military leave of absence.

Strengthening the company in all departments and divisions by the impact of their work, this loyal long-time service group is an invaluable bulwark in one of the most vital home-front industries.

Of the 1,124 employees with long-time records, 545 have between 10 and 20 years of service; 566 have between 20 and 30 years; 12 have between 30 and 40 years; and one, H. F. Miller, manager of the crude oil purchasing department, heads the list with 42 years of service.

The cross section of opinion among the Carter's veteran employees reveals that they have confidence in management, have found consideration and fairness in regard to employees, and feel that the company's program of employee benefits is exceptionally good.

Two More Carter Men Are Reported Killed

Reports of the deaths of two more Carter employees in the armed forces have been received. They are Lt. Walter B. Vincent, Jr., Tulsa, and Lt. Patrick H. Whittington, Shreveport, Louisiana.

These casualties bring to three the number of Carter employees

Lt. Vincent

reported dead in foreign operations. Lt. Harold J. Collis, former engineering clerk was reported killed in the South Pacific last October.

Lt. Whittington, 24, a former employee in the Carter's Shreveport offices, died of injuries in Italy on April 15, according to information received by his wife in a recent telegram from the Adjutant General:

Lt. Whittington left his duties with the Carter on July 29, 1942, to enter training as an Air Corps cadet. He began his employment in the Shreveport office on March 10, 1941, and had one year and two months service at the time he began his military training. Lt. Whittington was born October 28, 1920, in Shreveport, and attended public schools there.

Surviving are his wife and his mother, Mrs. P. H. Whittington, both of Shreveport, and two brothers.

Lt. Vincent, 21, former Carter Tulsa employee who was reported missing in action in the South Pacific one year ago, has been declared officially dead by the war department, according to information received by his parents, Mr. and Mrs. Walter B. Vincent, in Tulsa.

Carter's first employee to be reported missing in action in World War II, Lt. Vincent was navigator-bombardier on a B-25, and was lost in action over the New Hebrides group. He entered the Marine Corps in De-

Lt. Whittington

cember of 1942 and was sent to overseas duty in February, 1944.

Lt. Vincent began his employment with the company as a bookkeeper machine operator in the Carter's general accounting department on May 19, 1941. He was a former student at the University of Tulsa, and was a member of the first graduating class of Tulsa's Will Rogers high school. He was born December 11, 1922, in Bartlesville, Okla.

He is survived by his parents; a brother, Capt. Ernest Calvin Vincent, Quartermaster Corps, 15th Army headquarters; and two sisters, Mrs. Robert Washburn, Tulsa, and Mrs. Frederick J. Perry, Oklahoma City.

MILLER *(Continued)*

sampling of our products terminated my brief career with that little firm, and my ex-employer deplored my subsequent proposed acceptance of a position with the Standard Oil Company, at that time referred to abhorrently by such small business people as a "trust," a very wicked thing. Despite his forebodings, I took the chance of a business career "conceived in sin" and entered the company's employ by the office-

boy route at $18 a month, contrasted with $15 paid by my previous employer—an increase of 20 per cent.

Shorthand loomed as a means of increasing income, and I became a stenographer, or should I say, a "doctor of abbreviated chirography"?

Came later the opportunity of serving in a secretarial capacity to a top executive of our company, and there followed some ten years or more of annual and

semi-annual visits to the capitals and principal cities of Europe, journeys across the United States and into Mexico and several years sojourn in Canada. After that came increased responsibilities and twenty-four years of residence in Oklahoma.

I have not found anything for which I would exchange my experiences and, were it possible, I would like nothing better than to start all over again with our company.

THE LINK *for June, 1945*   Page 15

DEAR DAD

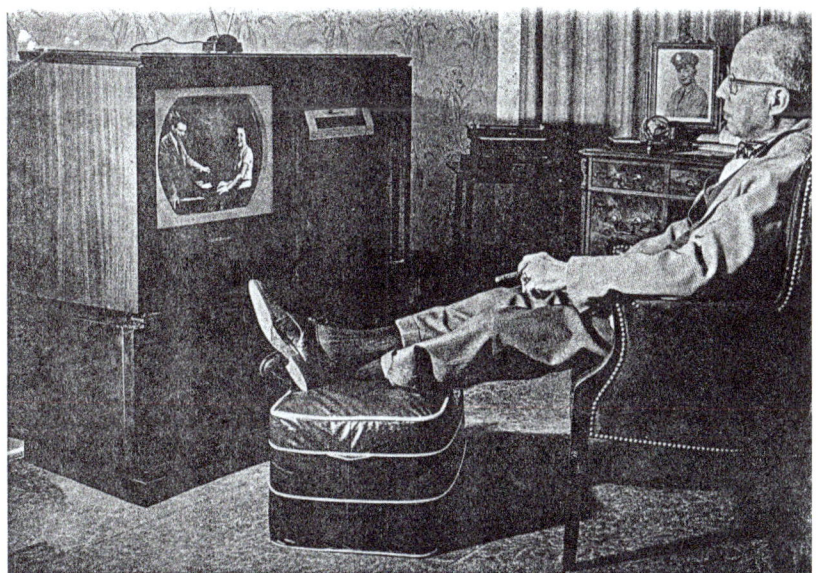

HERBERT F. MILLER RELAXES BEFORE HIS TELEVISION SET, A RETIREMENT GIFT PRESENTED BY A GROUP OF FRIENDS IN THE COMPANY.

Miller Retires After 48 Years With Company

Business Career of Boy
From Brooklyn Is Typical
American Success Story

It was 10 a.m. of a bleak and chilly February day, but the cold wind didn't bother Herb Miller. He was sleeping peacefully in his bed. Herbert F. Miller, vice president and director of The Carter Oil Company had broken with a 48-year work routine. He had retired.

Retirement February 1 brought to a close a continuous Company career which started more than 48 years ago when a 16-year-old Brooklyn lad took an $18 a month job as office boy with Standard Oil Company at 26 Broadway.

The story of Miller's climb up the business ladder, and finally retirement on a comfortable Company pension, could have been lifted intact from the pages of a Horatio Alger novel. The alert office boy was taught shorthand by an older man in the office. He became a stenographer. Next came an opportunity as secretary to Walter C. Teagle, former president of Standard Oil Company (New Jersey).

MORE THAN 80 WERE PRESENT AT THE DINNER HONORING MILLER ON COMPLETION OF MORE THAN 48 YEARS WITH THE COMPANY.

J. W. BRICE, Jersey director and former executive in The Carter Oil Company, visits with Miller about their many mutual friends before start of the dinner.

FRANK W. ABRAMS, chairman of the Jersey board, center, and Miller are long-time friends. Abrams recalled that Miller welcomed him to New York executive offices.

ABRAMS, left and O. C. Schorp, right, Carter president, give Miller their ideas on things to do and see.

J. R. FREEMAN, right, welcomes Miller into the ranks of the 200 annuitants of The Carter Oil Company.

The Brooklyn boy was on his way. With Teagle there followed 10 years or more of annual and semi-annual visits to capitals and principal cities of Europe and trips across the United States and into Mexico. Miller, with his world-wide experience in the oil industry, was drafted to head Carter's Crude Oil Purchasing Department. A year later he became a member of Carter's board of directors. Climaxing his Carter career he was elected a vice president in 1947.

Most persons never achieve 48 years with one company. When Miller joined Carter's 200 annuitants it obviously called for special recognition.

Frank W. Abrams, a long-time friend and Chairman of the Board of Standard Oil Company (New Jersey) headed the delegation from New York which came to Tulsa to honor Miller. Others attending were John W. Brice, Jersey

W. J. HALEY, President, Esso Export Corporation, served as toastmaster at the dinner and kept the guests laughing with his almost limitless supply of jokes and stories.

PRESENTING MILLER a "gold cane in recognition of 50 years of service," Haley sawed two inches off the end because Miller had completed only 48 years of work.

DEAR DAD

O. C. SCHORP, Carter president, tells an amusing story concerning Herb Miller and his career.

MILLER BOYHOOD scene was taken from a film shown guests which gave humorous highlights of his career.

A. H. MITCHELL, left, and F. W. Bruner enjoyed the Western music by the Sons of the Range orchestra.

Director; R. H. Sherman, Co-ordinator of Producing; H. W. Fisher, Co-ordinator of Refining, and W. J. Haley, President of Esso Export Corporation.

O. C. Schorp, Carter President, speaking at the retirement dinner said:

". . . The hard work, the intense interest, the ambitions, the mutual respect and appreciation of nearly a half century have gone into the building of the outstanding career we salute tonight.

"Carter and Jersey well know and appreciate the talents of Herb Miller. We know of his keen business sense, his devotion to his job, his intense and fully reciprocated loyalty to his company.

"We know him, too, as a good citizen, strong family man, a good neighbor to everyone everywhere, but best of all in a personal sense as a good fellow—as a great friend. . .", the President declared.

(CONTINUED ON PAGE 8)

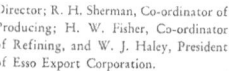
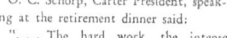
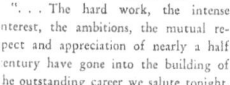

ABRAMS, left, Miller, J. J. Conry, retired Carter President, and Miller's son Frederic at the retirement party.

J. R. McWILLIAMS, Carter Executive Vice President, presented the television set which was a gift of friends.

H. W. FISHER, center, Jersey Co-ordinator of Refining, was in the New York group honoring Miller.

R. H. SHERMAN, left, Co-ordinator of Producing, with F. M. Darrough, R. W. Gemmer and O. D. Harper.

THE NATIONALLY famous Flying L Quartet and a Tulsa Opera Duo sang musical numbers during the dinner.

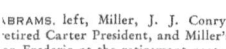
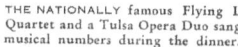

WORKERS IN CRUDE Oil Purchasing department were guests at a dinner honoring Miller on January 28 at Petroleum Inn. Ben L. Jones, speaking for fellow workers, told Miller he was the finest boss in the Company.

ABRAMS, in the principal speech of the evening, said:

". . . During the course of Herb Miller's career we have seen a lot of history written in this country. We have seen the increasing emergence of individuals generally as a vital factor in the growth and success of business. More and more we have seen management recognize the basic fact that the dignity of the individual is the real key to the maintenance of our competitive economy.

"You are all familiar with what we as a Company have tried to do in this direction, so I don't need to take up our program in any detail. I do want to say, however, that I firmly believe that—collectively—our Company represents as competent and enlightened an organization as this country or the world has seen. We hold strongly to the faith that there is no single employee who is not important to our continued development and success. In fact, we know that the Company's future depends to some degree on each member of its working family. And we are constantly doing everything we can to make it possible for him to help us progress and succeed."

Speaking extemporaneously, Abrams told the group that the day of quick riches in America is through. He said that doesn't mean that opportunity is lacking in America today. Abrams believes there still is tremendous opportunity for service and a full satisfying life in companies such as Jersey.

He cited as the best proof of this statement the Horatio Alger story climb of Herbert Miller and many others like him in American business today.

Abrams told the group he didn't know when Miller started out whether Herb planned it that way or not, but that it wasn't important. What is important is that he achieved that kind of life and earned the respect and admiration of all who knew him, Abrams said.

All employees of the Crude Oil Purchasing Department and their wives had honored Mr. and Mrs. Miller with gifts at a dinner party earlier in the week.

Ben L. Jones, speaking for the employees at that dinner, told Miller of the admiration and respect he had earned from all Carter workers. Jones concluded his talk by saying:

"You're the finest boss in The Carter Oil Company."

GUESTS SING and dance as Miller, famed for his lively piano playing, pounds out the latest popular dance hits.

C. M. YORK, long-time worker in the Crude Oil Purchasing department, presented gifts from employees to Miller.

Germany—October 2, 1944

Dear Mother, Dad & Kids,

I just got a letter dated August 25, I can't figure out why it took so long for it to get here. You remember that clipping you sent along with the one about Galloway and Bob Seaman—well you were right.

These Germans make me sick, every time we capture a town the civilians come out and say that they hate the Nazis and 5 minutes later one of them starts shooting at you and we have to shoot them. Right now we are in a town holding it. We have to keep a close check on civilians—check their identity cards, search houses, and stuff like that besides guarding the roads against the shells that come in day and night. I'm living in a beautiful house with my tanks just outside the door. These S.S. troops have a bad habit of sneaking in and dropping a grenade in your bed. It kind of tickled me to hear you talk of sleeping in Dad's room and locking the door while he was away. By some miracle the lights and water are still on—though I expect a shell to hit the power plant any minute. We've been pretty lucky lately about finding a good place to sleep. We just walk in a house and say "Rouse Mitten" or something like that which means get out in a hurry.

We are right near an old castle and it's really something—complete with dungeons and a moat and drawbridge. Underneath it's honeycombed with tunnels.

A lot of us have been sick lately—I guess it's this cold, *wet* weather and the same cold food day after day. It's getting so I almost get sick every time I look at a K-ration. I can't figure out about these cigarettes—back in the States they were always telling on the radio how many smokes they were sending over. Now we get 10 cigarettes per day if we are lucky and 10 of them don't go very far.

By the way, you said you didn't know what calvados is, it's a French drink that tastes like TNT but it sure helped on cold nights. In Germany we are afraid to drink anything because it might be poisoned. Please write soon.

Love, Jud

April 19, 2014

Dear Dad,
Now your battle sounds more like guerilla warfare and you needed eyes in the back of your head. So this is where you began to use the phrase "Rouse Mitten." I remember it well when you wanted us to get up! Another one you used often was when we whined. You would say, "Poor babies, was ist los?" No sympathy there.

I wonder what castle you were near. It sounded really interesting. With all those tunnels it would have been an ideal place for the enemy to hide.

I found a newspaper article and photo of Samps, Dave and Fred tending the family Victory Garden, which in this case, was a small chicken operation.

How different this war effort was in comparison to the way we wage war today. Now no one at home makes sacrifices during wartime. There is an unhealthy detachment at home, a distance that sanitizes our emotional involvement. We are spectators, not participants. We are delegating and do not want to dirty our hands. People mouth all kinds of platitudes as if everyone is a politician. There is such a lack of sincerity and honesty in their robot like clichés. I believe that we should reinstitute the draft. One would not necessarily have to serve in the military branches, but you would have to serve for 2 years in some capacity for the country. We would be better for it. For me the Peace Corps is calling loudly and soon I hope to go. I have a debt to pay. I also want to make you as proud of me, as I am of you.

Love, Your girl who hopes the Peace Corps will have her

October 7, 1944

Dear Mother, Dad & kids,
How do you like my new stationary? I picked it up yesterday.

Things were kind of dull in the town we are holding, so a couple of officers and myself decided we would look for a little excitement. We are right on the edge of the Siegfried Line so we went up to it looking for Krauts but instead we found a bunch of houses where German officers and their families lived—boy were they ritzy. They were full of whisky and a million other things. I took a beautiful radio and a pail full of bottles of whisky and champagne—we were afraid someone would come in. Just after we left they saw us and started shooting. You should have seen me running with that load. I was debating which to drop—the whisky or the radio, but I kept them both and we had a big party last night.

We are now allowed to tell where we have been up to within 25 miles of our present location so I'll start in naming the places I've been and later when I get a map I'll give you the names as I can't remember all of them—also my spelling might be a little off. I landed on the beach in a landing craft at Isingy, then went to Carrentan, Barneville, St. Marie Eglise, St. Lo, Vire, Domfront, Ambrieses Le Gran, Juvigny, Chartres, crossed the Seine below Paris near Melun and Corbeil, then to Soison (spelling?), Reims, Charleville-then to Belgium-Phillipville, crossed the Meuse near Dinant and Namur, then to Malmedy and that's as far as I can tell you. I left out a lot of places but when I get a map I'll remember what they were. I was very close to Paris but I didn't get a chance to go in. Charleville was really some town—we really had a big time there, the people just about mobbed us they were so glad to see us. I ate dinner with a man there and he said he would write you. He really has a swell place. He owns a factory that makes window glass—he said he would really clean up after the war.

By the way, in addition to the radio and whiskey I found a small camera yesterday. I didn't find film but I think maybe I can get some.

This Siegfried line is no joke, it's plenty rough but we'll get through it alright. But I'm afraid this war is going to last *much* longer than most people think, and after its over I'm sure I'll either go to the Pacific or stay in army of occupation and it looks more like I'll go to the Pacific. Anyway I know I won't be sent home. The only thing I like about this country is the beer which only costs 2 cents per glass. In one place I lived and slept in a café (bar) for 8 days.

Please save copies of Life as I've seen pictures of several places I've been. I saw one picture of the beach where I had been. I've been paid and will send $140 home as soon as I get the money order. Please write soon.

Love, Jud

April 21, 2014

Dear Dad,

What kind of war was it? I would give anything to have a picture of you racing away from those "ritzy" houses clutching your ill-gotten gains of whiskey and champagne as if your life depended upon it!! What a laugh!

I never told you this, but when we lived in Bad Kissingen a bunch of us went into a farmer's field and were eating his strawberries when he unexpectedly showed up. He started yelling bad German words as he ran at us shaking some kind of a deadly looking farmer tool. I know I heard him say "scheiche kopfs." Boy, were we scared! Those strawberries sure tasted good. We never went back and as far as I know no one told their parents either. I also must confess to the "disappearance" of a bucket of Peter Pan Peanut Butter around the same time. Mother had gone shopping and I took the brand new bucket along with a spoon and made my way to the creek in back of the duplexes. I commenced to eat the entire bucket. I then had an attack of extreme elimination of said peanut butter both orally and through the other orifice of the digestive track. To this day I am not very interested in peanut butter. To add insult to injury not too long after that episode I became angry about something and told Mother that I was running away from home. She dutifully packed peanut butter and jelly sandwiches for me, wishing me the best on my journey. I left and went to the edge of the yard until dark before finally giving up my quest for a better life outside of the family. How could I go on with only that dreadful peanut butter to sustain me?

Now let's fast forward to your grandson, Travis, at age 3. Travis loved peanut butter and jelly sandwiches. He would chant this little sing song thing that musically rose higher and higher, "Peanut butter n jelly, peanut butter n jelly, peanut butter n jelly…" until whoever was available would break down and make him one. One time Liberty, along with step-brother Sam, decided they had their fill of this performance from said "little bratty brother." They made him a peanut butter and jelly sandwich and laced it with hot sauce. Needless to say, when I returned home Travis showed me his scorched mouth and tongue. He was sure that he would lose the ability to eat as well as taste as a result of this crime against "the bratty little brother." It seems that we have a pernicious criminal streak in the family.

I am fascinated to learn where you have been. You told me once you knew Dick Winters. I can only imagine how you met and what the circumstances were.

Love, Your girl with the peanut butter phobia

P.S. I need to whine a little. My lawnmower has given me a medical condition. I had the thing serviced just as you are supposed to and mowed once. The second time it refused to start. I pulled that cord and pulled that cord. I cursed it with all my might to no avail. The entire time I was thinking, "Kathy you should have dumped last year's gas." It is always a real treat to observe my irrational side. Just get mad at an inanimate object and see what it gets you, a medical condition and more expense at the mower shop to get rid of the old gas in the carburetor. Now I have a crick in my neck and I am grouchy. On top of that I am pretty sure the cat had a smirk on her face as she watched my useless efforts from under the gardenias. She probably went and told all her cat friends about her stupid human who makes all kinds of grunting screaming noises while performing a weird ritual over an object that does not know how to groom itself or perform any useful cat functions. Whine, whine poor baby.

Germany—October 9, 1944

Dear Mother, Dad & kids,

Just a note to let you know I am okay. I'm enclosing a 10,000 mark note made in 1922—it's no good now because it was inflation money—if it was good it would be worth $1000. I nearly fell over when I got it.

The weather is the same wet and cold. I found a pair of skis and when it starts snowing I will give them a try.

I got the paper today and really enjoyed it. I've read it about 4 times already.

I don't have my money order yet but as soon as I get it I'll send it home. How much more do I have to go on the car? It seems funny to have to pay for a car when we shoot up 30 or 40 Nazi trucks costing several hundred thousand bucks. Our news broadcasts are just as full of propaganda as the Krauts—I know for a fact some of the stuff are downright lies. That's all the time for now. Write often.

Love to all, Jud

April 22, 2014

Dear Dad,

The obsession with the car continued no matter how surreal your circumstances were. I wish I could find out what kind of car it was. Perhaps Aunt Betty knows. I will be seeing her in June. Shelley, Doug, and I are going to Tulsa in June to attend the OMA reunion. Since I am still doing research for this book I thought it would be a good idea to go there and tour the museum where you are in the Hall of

Fame. When I contacted them they invited us to come to the reunion. They have made a TV documentary, *Oklahoma Military Academy: West Point of the Southwest.* It will premiere at the reunion. I am hoping to meet some of your fellow classmates. To prepare for this event I did some research. This means I read the comments in your yearbook, the Vedette 1942. Now I really have some eyebrow raising questions for you.

First just exactly what is a "T Town Buddy?" Many of the comments from your classmates refer to you as their "T Town Buddy." Does this refer to the fact you were from Tulsa? Then there are many references to "The Gay Boys of The Thirsty Three." Back then "gay" did not describe your sexuality but it seems to me that you were part of a drinking club of some kind. Comments included, "The Thirsty Three Forever," "Thanks for the "pass," To a Pig Stand Buddy." "Hot Lips" (this turned out to be the champion bugle blower, John De Tar, not a babe like Hot Lips Hoolihan in *Mash)*. "Hot Lips" wrote extensively about how when you were on guard duty you would wake him up to come to the guard shack to imbibe the barrel of "beah" that you had there. What the heck does "Take it away Leon" mean?

However the crowning glory of this yearbook of salacious innuendo and tantalizing tidbits has to be what the school nurses wrote to you. You made quite an impression on them as preserved forever in printed memory. They both were "babes" as you liked to describe hot women. What I also found so intriguing was that Mother, who you would meet at the end of the war, was also a nurse. In the comments of Griff is the first mention of the song, "Take Me Back to Tulsa." I wonder how many times you sang that song to us. "Take me back to Tulsa; I'm too young to marry…" Ha, Ha!!! My father, hitting on school nurses at the tender age of 18. What else did you communicate to my son, Travis, when I was not around??

Love,
Your girl who plunders Dad's old yearbooks for tantalizing tidbits

Germany—October 12

Dear Mother, Dad & Kids,

Enclosed are money orders totaling $140, please let me know if you get them. I figure now I've sent home about $420, let me know if that is right and how much I still owe on the car.

I'll tell you a couple of things I could use—a good fountain pen, this one leaks and doesn't work half the time, a metal watch band—I've got my watch tied on with a string now.

There isn't any more time so I'll close for now. In this town we are holding you can get a shave & haircut for 1 mark (10 cents), so I get a shave every day.

Love, Jud

Germany—October 22, 1944

Dear Mother, Dad & kids,

Nothing much to write but I thought I'd let you know I am okay. It looks as if this war is going to last a long long time, because this Siegfried Line or "Westwall" as the Germans call it, is certainly no joke. I'm living pretty well now in a house that adjoins a café (beer joint). All we have to do is hold the town against counterattacks. The Krauts shell us all the time but it's more of a nuisance than anything else.

Today they started giving furloughs to Paris but it will probably be a long time before my turn comes up. I'm sure looking forward to it.

What is really spooky is walking around town after dark in the blackout. I'm always afraid some civilian will jump out of a doorway.

I'm really getting where I can speak German quite well. I get a big kick out of talking to civilians, they all tell the same story—they always hated Hitler, didn't want to fight, and so on, I'm not dumb enough to believe that. The Germans have really had heavy casualties—every time you ask a civilian "Vo ish eer fatter (brudder) (mahn)—let Dad translate—they say he is "Kapoot" (dead). Incidentally that word is becoming part of our slang. My fountain pen is Kapoot. Kapoot means dead or finished.

The other day I managed to leave the front and go back to a very large city in Belgium to buy some clothes at the quartermaster in Vierviers. It was very expensive and wasn't much good.

I've seen and heard flying bombs but they are more a nuisance than anything else. They sure make a racket when they go over.

I was caught in an air raid in Vierviers—such confusion you never saw, we went into a shelter and was it packed. Another thing I don't like is the blackout, you have to have a map to find your way home.

So Fred, you want to know what a K-ration is. It consists of 4 crackers or biscuits (hard as rock), a tiny can of hash or cheese, 3 fags, a stick of gum, a piece of candy, and soluble coffee all in a box about 8 inches by 2 inches by 3 inches and believe me it's awful.

Please write often because your letters are really appreciated. By the way, I made an allotment for $140 this last month and Uncle Sam should send you the check in about 6 weeks. I think after this though I'll use money orders.

Love to all, Jud

April 26, 2014

Dear Dad,

This lull in the fighting seemed to have you a little bored. Once again it is interesting to observe an attempt to have some normalcy in the midst of war with the availability of the 10-cent shave and haircut.

It was very convenient that you chose to set up housekeeping in a house fixed to a beer joint. Did you guys have any rules as to when you were allowed to drink? Was this set up on a shift schedule? Let's see, ok I will take the 3-11 drinking shift and you can have the 11-7 shift.

I have never experienced anything remotely similar to a blackout or been worried about people suddenly appearing out of doorways. I did have one creepy experience that may have exacted the same sort of "hair on the back of your neck rising thing." I had just moved to a new home in Waycross and this move traumatized the cat. Yes, I know, another cat story but just bear with me for a minute, please. Gracie stayed hidden for about a month. She must have come out while I was at work or while I slept but the damn cat hid from me for a month. I was seriously debating getting a cat psychiatrist for her. One day I came home from work, went into the kitchen when all of a sudden I felt someone looking at me. I tried to remember if I had locked the back door. I broke out in a sweat and my head got so hot my hair fell. As I furtively looked around for a weapon I let out a shriek as I saw two eyes glaring at me from between the cabinet above the refrigerator and the ceiling. Panic mixed with relief as I realized that I was looking at the "move-traumatized cat." As I walked back and forth in the kitchen she moved her head back and forth as if to stalk me for her next meal. All of a sudden this feline beast that I had not laid eyes on for a month leaped from her lair out onto the floor and made a mad dash for parts unknown. It was so sudden that I shrieked again, much to my embarrassment. Damn cat.

I remember "Kapoot" from early on and you are right, even today it is part of our language. How fascinating to learn that you could hear the flying bombs, the V-1, as they flew over you. The history that you have been part of is truly remarkable. I remember the first time I learned that you had witnessed the explosion of an atomic bomb. We were in Las Vegas and you said that you did not like Vegas then any more than you did the first time you were there. When I asked when that was you stated casually, "Oh, I was out here watching an atomic bomb go off." Holy cow, Dad! Did you realize how amazing that was? How incredible your life has been to be part of history between the V-1 and the A Bomb?

      Love, Your girl who shrieks in fear from house cat

Page 5-date and beginning gone

 I got a swell officer's field coat (waterproof & warm) price $30 but it was worth it and a pair of tank boots. It was the first time I'd been out of German artillery range in 3 months. I saw some nurses—the first American women I've seen since I left England. It was almost like a holiday to go back to the rear. This cigarette deal makes me sick. The damn niggers in the quartermaster have so many smokes they give them to civilians and I've been smoking Kraut cigarettes from dead Germans. Boy are they lousy. Up here I've paid as high as $2.00 for one package. I'd appreciate it if you would mail a carton per week and take the money from what I send home.

 I like the German people pretty well on the whole, much better in fact than those lousy English. The Germans would be okay if they would get someone to run this country. Everything is clean and much more modern than France and England. As far as I'm concerned we can lick the Limeys after we finish the Krauts.

 I found another very expensive German camera with a lot of film so if the weather ever clears up I'll start taking pictures. I'm all set for the winter now with a pair of good skies I found (my tank is beginning to look like a junk wagon). Please write soon and don't worry,

                Love to all, Jud

April 26, 2014 PM

Dear Dad,

I do not know where the rest of this letter is so I am guessing at the date but since it details your leave I think I have it pretty much in chronological order.

You used some bad words. The worst I ever head you say was damn and damn it. I never heard you use the word "nigger." Not ever. In fact I never heard the word until we moved to Tampa when I was 13. I must have had a sheltered life growing up with mainly military families around me. The closest thing to racism that I can recall was how I felt about "civies" and that bordered more on military snobbery towards civilians than racism. I had never lived in the South until then. I never heard you talk or say anything remotely racist about anyone. Perhaps you were using this more as a figure of speech. In fact I remember a story you told me about the brother of Lena Horne who was at Ft Lewis the same time you were. You held him in high regard and respected his professionalism and integrity.

You were pretty tough on the English; I can understand your admiration for German cleanliness, order and efficiency, even if they had Hitler. What is that saying? People usually get the government they deserve? Perhaps "Evil thrives when good men do nothing?" Mimi has drawn lines through some of the letter. She would type copies of your letters home and mail them to family so I think she was using the lines to censor anything she felt might make someone mad. It is interesting to compare what she typed with your originals as they do not always have the more controversial parts left in them.

Love, Your girl who is probably somewhat of a military snob

Victory Garden

Smirking cat

MEDICAL STAFF

DR. P. S. ANDERSON
Academy Physician

DR. J. C. BUSHYHEAD
Chief of the Medical Staff

The Hospital

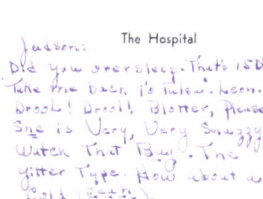

BERTHA GRIFFITH
Assistant Nurse

KATHERINE TAYLOR
Head Nurse

Yearbook tells a tale

NOVEMBER 1944

Germany—Nov 4, 1944

Dear Mother, Dad & kids,

    Sorry I haven't written in so long but I just haven't had time to do it. I just returned from a three-day pass in Vierviers, Belgium and really had a swell time. It's the first time I've been away from the front since I joined the outfit and it sure was nice not to hear any shooting. Another officer (from Okla. City) who is in this outfit went with me. I had my first ice cream in 6 months and ate a ton of it. About all we did was eat ice cream, drink champagne, and sleep. These cafés are really something. Those people really soak you though, one jigger of cognac costs about $2.00 and in 20 minutes we ate $6.00 worth of ice cream, but it was worth it. They even had an officer's club and last night they had a dance with U.S. nurses, WACs and a bunch of Belgian girls, it was pretty nice but we combat troops with our beat up uniforms and muddy boots didn't stand a chance with the Quartermaster and Air Corps officers in their pinks. Those guys sure make me sick—they complain because they have to work Saturdays, if I could get one of them up here for a day he would be glad to go back and work 24 hours a day. Gosh, what an easy life they have.

The lens on my German camera broke but I got a few snapshots first and I'll send them home this week, also the camera.

No packages have arrived yet but I guess it takes a long time. I tried to find something to send home for Xmas but everything....

{REMAINDER OF LETTER MISSING}

April 27, 2014

Dear Dad,
Unfortunately the last part of this letter is missing. Still, I am able to enjoy reading about your brief time away from battle and your FFDF, gluttonous, ice cream bacchanalia.

Our entire family has a deep love of ice cream. I can certainly understand your ravenous consumption when the opportunity presented itself even if it is FFDF. We especially love chocolate chip mint ice cream. Every summer when Travis and I would visit you in Tacoma we would hit the ice cream parlors with a vengeance. I remember the first time you had a blizzard. We were on our way back from the mountain and stopped at a Dairy Queen in Puyallup. Travis and I both ordered a Heath Blizzard and you, ever curious about new things, tried one as well. It started a new tradition when we would visit Dairy Queens, the Heath Blizzard ruled!

You would always stock the freezer full of chocolate chip mint ice cream as well as some Haagen Dazs. Every night you and Travis performed "The Ritual of Proper Ice Cream Consumption."

Each morning Travis would head straight to the freezer for a nourishing bowl of chocolate chip mint ice cream for his pre-breakfast warm up. Every morning someone had depleted the chocolate chip mint stock to the point that his bowl was never as full as he desired. At times there was even the ultimate tragedy, no chocolate chip mint ice cream at all! This, of course, necessitated a mission to resupply the stock. Who was devouring this important food source? Did we have a Goldilocks visiting in the dead of night? Travis was determined to catch the culprit in the act but stopped short of guard duty outside the freezer. He never did identify the ice cream napper but instead put his own dent in the supply as he made several middle of the night reconnaissance trips to the freezer.

One time, as we were watching the evening news together, there was a story about the execution of Timothy McVeigh, the Oklahoma City Bomber. At the end of the piece the newscaster listed the menu choices for his last meal. The last item listed on the menu was chocolate chip mint ice cream! You turned to me and said, "Well, he couldn't have been all bad." Ha. Ha! I so loved your sense of humor! I really miss it too, Dad. That simple statement, aside from the humor, also taught a lesson. No one is totally immoral and there is always the possibility that we have something in common, something we can share.

Love, Your girl who did not midnight requisition chocolate chip-mint ice cream

Germany—Nov 13, 1944

Dear Mother, Dad & kids,
   Well how is the weather back home, over here there is more snow on the ground than I ever saw in Okla. and it looks like it has just begun. If it gets much colder I don't know what I'll do. Right now I have on long underwear, woolen shirt & pants, coveralls, sweater, combat suit (that's a windproof & waterproof-ha ha-suit, a jacket, overcoat, and wool cap and helmet and I'm still cold. A few nights back something interesting happened. I was on guard (even officers stand guard over here) and it was dark as pitch. I saw someone walking about 10 yards from the tank. I thought it was one of my men so I told him to get back to bed and he answered, "Var ist" or something like that so I let go with my tommy gun. After I shot him he said he was a Frenchman and started crawling towards me but I could hear him dragging his rifle so I finished him off. At night these darn Krauts are all around you. They sure are fighting better and harder since we hit Germany. We have a quaint way of enforcing the blackout for German civilians—when we see a light we just cut loose with a machine gun. I wouldn't trust one of them. We have caught 12 year old kids of the "Hitler Youth" sniping at us. That guy I told you about earlier in the letter was certainly no Frenchman—he was a full-fledged Kraut.
   Here are some pictures I took with my Kraut camera. I broke the lens on it so that's all I can take till I get the one you sent. One picture is of me by my tank—the other one is me by a pillbox I blew up. Please write soon.

Love, Jud

May 6, 2014

Dear Dad,
   You look grim. Very, very grim, especially in front of your tank.

You don't look young any more. Just grim and imposing. There is a trace of cockiness there as well, a look that dares someone to bring it on. I really like these two pictures of you.

My life has been hectic lately with finals and store inspections. I worked very hard in my Spanish class so I could exempt the final and thank God I did, because the professor gives really super hard tests. I can only imagine what his final was like. The only chance we had for extra credit was to cook some Hispanic food and bring it to class on the last day. I made empanadas for the first time and, let me tell you, they were awful. Probably as bad as those K Rations you are eating. They were too dry and the dough was too thick but I got the extra credit. I bet he changes his mind about that kind of extra credit next semester.

I had my research methods class final today. First I need to tell you that my professor, Dr. Darin Van Tassell, is allowing me to work on this book for the methods class as well as for Senior Seminar in the fall. How many professors would do that for you? Secondly, we had all studied very hard to prepare for his class final exam. We had been given themes and would have to write essays in class. However when we got to class he handed us the same copy we had previously received to study but this one had another page; since we were so well-prepared we only had to write a very different essay starting with what grade we thought we had earned. It blew my mind!

I can hardly believe that I will finally get my degree in International Studies in December. I so wish you could have been here to see it happen. I will also have a minor in Spanish since at Georgia Southern a minor in a foreign language is a requirement for the International Studies degree. I have been at this for 14 years; 15 years if I include the one-year when I was at USF at age 18. You also went to college for about the same amount of time. I think you told me it took you 15 years. I often thought about you when I became overwhelmed with work and college. It helped sustain me and while you were alive you constantly told me how proud you were of me as I pursued my studies. I really miss that. I really miss calling you every night at 8 pm and getting to hear your voice.

> Love, Your girl who misses her Dad's voice

Germany—Nov 25, 1944

Dear Mother Dad & kids,

   I've been getting quite a few letters from you lately but as you've probably seen in the papers things are pretty active over here and since I'm right in the middle of it I don't have much time to write.

   I'm enclosing 2 postcards of a town we fought in for quite a while. The name of it is Monchau and it is southeast of Achen. Thanks a lot for the War Bond, I really appreciate it.

   I wish we were back in France fighting—that was almost fun compared to this place. I have not shaved, washed, or had my shoes or clothing off (even at night) for two weeks. It rains every day and I don't mean maybe—the only time it stops is when it snows. The mud is so deep it almost swallows you up and on top of that the Krauts are fighting harder than ever. I'm afraid this war may last well into next summer. The Krauts don't let us surround them like we did in France. They just fall back a few hundred yards and we have to start all over again. Our Air Corps can't help us because of the weather so we really have it rough. That Air Corps is the racket—no mud, no rain, a warm bed at night, and extra pay to boot. Cigarettes are still scarce, I traded my Lugar for a carton of Chelsa cigarettes and I could have sold it for $100 in Vierviers.

   I haven't received any packages yet but I'm still hoping.

   Please write soon and thanks again for the War Bond, it really means a lot. How is the West Point deal coming?

> Love to all, Jud

May 10, 2014

Dear Dad,

Funny how the mind works. I read your letter and I imagined you not bathing for two weeks, which for you, must have been pretty dreadful because I know you were very fastidious. Bathing can be such a cleansing, relaxing, and invigorating experience. But in my consciousness, a bathtub is a horrific memory.

I remember when Juddy was born, late in November at Ft Leavenworth. He was the epitome of you. It was as if you had been cloned; the two of you were so very handsome with brilliant blue eyes and blond hair. Ha! The two of you could have made a perfect Nazi Germany photo; not too surprising considering our German ancestry. Dougie and I were the two brunettes, Shelley and Juddy were the blond wonders!

This particular day we were living in Germany; I was probably 8 or 9. Shelley and I were playing out back; you were asleep in the chair downstairs while Mother was rounding up Dougie and Juddy to give them their bath. It was late afternoon and the weather was fine. I am pretty sure it was a Saturday. I heard Juddy crying but he was a baby and he cried a lot as babies do.

There was a terrible, anguished scream. I had never heard anyone scream like that. I could not even tell who was doing the screaming. I ran into the house to see Mother racing down the stairs with Juddy wrapped up. You had leaped to your feet and as you looked at the wrapped bundle that was Juddy you cried, "Not burns, not burns!" I had never seen you cry, I had never seen my Mother like that, I had never seen anyone like that.

Somehow Juddy had been left alone for a few seconds. He had turned the hot water on and suffered third degree burns on his legs and feet. I remember Mother had complained before the hot water was too hot and the thermostat needed to be turned down. Too late, too late.

Mother and Juddy left Germany and went to Walter Reed in the States for 6 months. We had several German maids that cared for us.

There were whispers that Juddy had a 50/50 chance of dying. People did not know that I heard them; that I understood. I would lay awake at night terrified, asthmatic and helpless.

Mother and Juddy returned. Before the accident he had been walking. Now he had to learn how to walk all over again. He did. He walked in spite of the hideousness of his burns that made his legs look like melted wax, only horrible and still painful at this point for him. If you ever watched *Raiders of the Lost Ark* and saw the face melt on the bad German Nazi guy, that was what his legs looked like to me. Mother still had to change bandages on him. He had started to talk while he was gone. An orderly at Walter Reed had taught him to say, "Hot dog Buddy, Buddy." He said that a lot after he came back. We loved to hear it. Sometimes when he would say it I had to run away and hide so I could cry. His legs were ruined but he could say, "Hot dog Buddy, Buddy."

We were never the same after this. Mother blamed herself. You blamed her. Others who liked to gossip and say mean things about other people would say that she was a "negligent mother." I heard them all. I saw it all. How I hated them. They would make comments about Juddy, too. Such rage I held inside of me. My Mother loved us, she was not negligent. At the time of the accident she was an officer's wife with 4 children ages 18 months to 9. We were a handful and you were gone a lot. While we lived in Germany Shelley caught the measles and lost her hearing in one ear. Dougie had to have hernia surgery, something very scary back then for a small child. But Juddy's accident destroyed her. Oh how she suffered. I would hear her crying in the night when she thought no one heard. But I heard her. Over the years she would deteriorate. She never forgave herself. To the very end of her life she cried for her baby Juddy. Her slide into mental illness was wrenching for me to watch. I was helpless. I became the "Little Mother." She needed mothering. Back then depression, thoughts of suicide, and any evidence of mental illness, was treated like a dirty little secret that no one talked about except in whispers. Her madness made sense to me. We grew up with our crazy but loving Mother. We always knew she loved us and desperately needed us.

You were also deeply distraught. You did blame her. I tried to hate you for that. But ever the soldier, you kept your emotions in check and soldiered on with great emotional self- discipline. There was always a formality about you, a protocol. Emotional stuff was not allowed to spill out, no matter how overwhelming. But I was ever aware of your suffering as well. I saw through the strict self-discipline.

You and I never really had an honest and raw conversation about Juddy. Later in your life we did have several about Mother. We had an unacknowledged understanding. You and I could never talk about things that were too emotional or painful. I did not want to talk about it and neither did you. Then it got even worse. Juddy shot himself with a 12-gauge shotgun at age 25, a bad romance and final exams. His suicide letter told Mother he did not blame her for his legs. He never talked to anyone either. Gone at age 25. I hate thinking about everything he has missed. The one time I went to his graveside with you I saw you crying. I decided that I was never going there again. This is why I did not go back, I could not bear to see you cry, Dad. I hope you understood that. I probably should have told you before now but somehow I think you understood.

Love,
Your daughter conditioned to be Little Mother who understood

Germany—Nov 29, 1944

Dear Mother, Dad & kids,

I got your letters today dated October 29 and another one a couple of days ago. I sure enjoy hearing from you all since mail call is the *only* thing we have to look forward to except the end of the war and that looks like a long way off. I'm sure glad you enjoyed your trip back East. As for me, I'm getting a little tired of traveling—this tour of the world is anything but fun. I wish it would end.

I hear smokes are getting scarce in the States—I wonder who is getting all of them—I think the English are.

It sure makes me mad to read about 4-F's striking and quitting war jobs. I've seen American soldiers killed just because our artillery didn't have enough shells. When we get back those guys better keep out of sight because a lot of these Yanks would just as soon kill them as not.

We've had a couple of clear but cold days but it's started to rain and freeze. I've got on enough clothes for 5 men and I'm still cold. Right now we are in a big forest—I'll be glad when we fight our way out to some houses again. Nothing gives me more pleasure than to kick a Kraut out of his house and move in. We used to just make them move into part of the house but we found too many spies who would do anything to get you. It serves them right anyway. In Belgium and France the Germans had killed everyone in a village and civilians with their tongues pulled out and finger nails pulled out by the S.S. I sure wish we were back in France fighting—boy the people just swarmed over you. I'll never forget one town in Belgium, we were still fighting in the main street and bullets flying all over when an old man came running out yelling, "Viva La America" and climbed up on my tank with a bucket full of champagne. There we were—one hand on my machine gun, the other holding up a bottle of champagne. In Germany all you get is a dirty look and a shot in the back if you are not careful.

Well my birthday and Xmas will be here pretty soon but I guess it will just be like any other day except that it will probably be a lot colder.

I think tomorrow if I can find time I'll send a box full of souvenirs home. I can't send guns but I have a lot of Nazi insignia and other junk. That's all for now, please write often.

Love, Jud

P.S. Please go ahead and buy war bonds with all the money I send home.

May 12, 2014

Dear Dad,
It must have been hard to read about your folks traveling back East as if all was well with the world. But to your credit there was no hint of envy or resentment. I know now that at this point in the war you were fighting in Hurtgen Forrest and it was a terrible battle. The trees would explode and impale soldiers. I am really glad you had a tank. Growing up we would go and climb into every tank you could find for us. Parades, reviewing the troops, and climbing on military equipment were our normal.

I did not know that 4-Fs were striking and quitting. That would seem like treason to me. I wonder what became of the old man in Belgium who lauded you with champagne. It sounds like something out of a Hollywood movie! There you are blasting away with one hand and champagne in the other! Well done, Dad!! I'll have to drink to that!!

Liberty is getting married this week. I wish you could have met her fiancé, Ryan. They seem very happy and most of the family will be there. I will send you some pictures as soon as I get them. Doug is coming up here and riding with me to the wedding.

Love, Your girl who toasts for gun blasting,
champagne drinking Dad

Dad & Travis with correct ice cream spoon placement

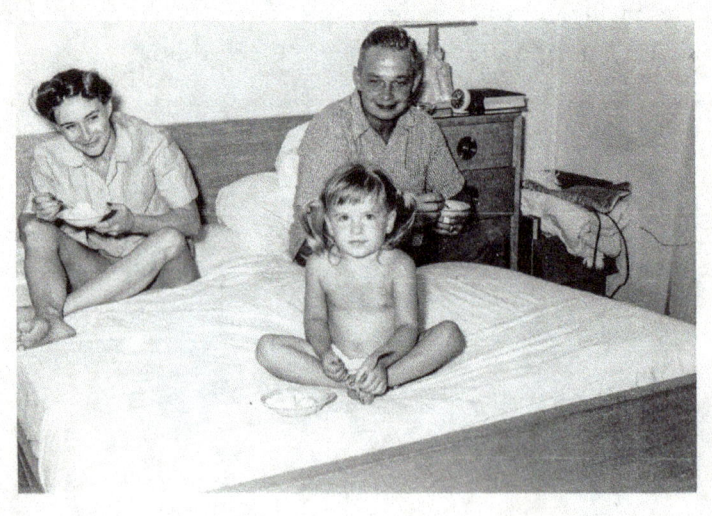

Mother, Kathy and Dad-ice cream tradition

Dad & his tank

Dad & blown up pillbox

Dad and Juddy Bad Kissengen, Spring 1960

DECEMBER 1944

Left to right: Good Buffalo-Indian and assistant driver, Scardella-Italian and gunner, Dad, Kush-Polish and tank driver

Germany—December 4, 1944

Dear Mother, Dad & kids,

If this paper is a little muddy you'll have to excuse it because everything I have is covered with this lovely German mud.

Well tomorrow I'll be 20 yrs. Old—doesn't look like I'll have much of a party. Something must be wrong with the mail because I still haven't received a single package, but some of the men are receiving packages mailed as long as 6 months ago so I'll probably get mine sooner or later.

I didn't send this month's pay home yet because I'm hoping to get a pass to Paris someday soon and I'd hate to have to turn it down because I was broke. So I'm going to keep some money on me from now on till I get the pass. At the end of this month I'll send home my pay for Dec. and you can put it into War Bonds along with the rest. If this war keeps going like it is I'll be rich by the time I get home.

Most of the Krauts we capture are pretty good soldiers but the other day we captured a two man bazooka team consisting of a man of 55 and a boy 9 years old—what a pair they made.

I broke the strap on my watch the other day and lost it. I sure hated to lose it as I've had it so long and it always worked good. I got a watch off a Kraut and I'm wearing it now but it's not much good.

I'll never complain about a hard mattress when I get home after sleeping in some of the places I have recently. I'm now in a forest where there are no houses and the worst part about it is that when Kraut artillery comes in it explodes in the tops of the trees causing what is called a "tree burst" which is much more dangerous than if it had hit on the ground. Consequently we have to either sleep in the tank or build a foxhole with a roof on it which requires about 6 hours hard work. Neither place is very comfortable.

It now gets pitch dark by 4:30 as the sun doesn't come up (actually we never see it) till about 8:00 in the morning which makes the nights awfully long and the days short. We eat only two meals a day and since there is nothing else to do we crawl into bed at 4:30 and lie there till we go to sleep. I sure will be glad when we get back into a populated area. Once I get out of this forest I never want to see another tree.

Do you remember an officer named Thompson that I brought down from Gruber once? He joined this outfit the other day. I certainly was glad to see him. Well that's about all for now—write often.

Love to all, Jud

May 13, 2014

Dear Dad,

Things must have been getting desperate for the Germans if a 55-year-old man and a boy of 9 complete a bazooka team. I think about my own son when he was 9. It is just inconceivable to picture him with a bazooka trying to kill someone.

Uncle Fred told me a story that happened when you were not much older than 9. It seems you were very adventurous. Fred said, "Your Dad tried to kill me and Dave once but did not succeed." You convinced Fred and Dave to go on an exploration. This adventure consisted of a hike through the woods until the three of you came to the Arkansas River in Tulsa, not far from your house. There was a railroad bridge over the river. You took the lead and set out for the other side with Dave and Fred in tow. You had chosen what you believed a safe route, under the part of the bridge where the train would travel. What you did not foresee was the violent shaking that the bridge would experience if a train should travel while you three undaunted explorers made your way across. Also there was not a footbridge, only some beams and crossties to cling to. A train appeared while the three of you were about halfway across. Fred told us he was convinced that he was going to be heaved into the river below at any moment. He was certain of his impending death. Fortunately the three of you clung to life and eventually made your way back to the riverbank. Fred said that you made them swear some kind of a forever oath never to reveal the details of this near death experience, especially to Samps. Death was to follow anyone who violated this sacred oath. Fred promptly violated the oath as soon as he reached the safety of home and you were in big trouble.

Fred even took me to this bridge and I was quite impressed with the chutzpa it would take to attempt such an endeavor. I also remember asking Mimi why you started at OMA in the 10th grade. She told me she and Samps could not do anything with you and they hoped the Army could. Ha! Ha!! It seems to me that war gave you plenty of adventure.

Love,
Your girl who was well behaved when she was young, sort of

P.S. When I was just six weeks from graduating from high school and you were in Vietnam, I came home drunk one night and Mother threw me out. So I took the one and only family car and slept in it for several nights on the Causeway. Then I found a part-time job earning just enough money to pay rent for an apartment. I graduated, took the car with some of my hippie friends and headed to New York City for an adventure. It was very selfish of me to take the only transportation that Mother and the kids had. By fall I had returned and started college and at that time you somehow communicated with me that the car really needed to be returned. The whole summer I had the credit card and either you or Mother paid the bills. I am not really sure how that worked. I did return the car. The crazy thing is you never, ever scolded me for my transgressions. I kept waiting for some kind of recrimination but it never came. Later that fall you returned to Tampa and when we met you were so kind to me. I did not deserve it but that is what you gave me… unconditional love and understanding. I love you Daddy.

Your hippie daughter

Germany—Dec. 5, 1944

Dear Mother, Dad & kids,

    Just a short note to let you know I got your letter dated Nov. 15. I also got 2 letters from Patti and a package from the Aches so my birthday was pretty good with all those letters and the package. Tell the Aches I really appreciated the package and if you'll send their address I'll thank them myself. I guess the packages will start coming in now I hope. I made myself a beautiful foxhole today. I dug a hole about 4 ft. deep and seven ft. long and covered it with logs and dirt. I got a battery and rigged up an electric light—it's really swell. Tonight I'm lying here writing, reading, and eating while shells are falling all around but I just laugh at them from my foxhole deluxe.

    That clipping you sent of Evelyn Davis was another Evelyn Davis or else she has a new face.

    Please keep sending as many packages you can as I'm sure they'll catch up and I'll really appreciate them.

    Yes it would be nice if the war were over by Xmas but seriously the way things look now I would not be surprised if we were still fighting next Xmas. Their whole army, including the air force, is **very** much stronger lately. In my opinion (and I ought to know) much of their equipment is just as good if not better than ours and they sure know how to use it.

    Well best wishes for a Merry Christmas and a Happy New Year, I'll be thinking of that big turkey and I hope next year I'll be with you.

Love to all, Jud

May 15, 2014

Dear Dad,

    I am very impressed with the fact you dug a foxhole 4 feet deep and 7 feet across. How long did that take you? Did each man have his own foxhole or did you share? The ground had to be hard and frozen but the background noise of artillery rounds would be motivation and an incentive to dig hard, fast, and deep. I wish there was a picture of your "Foxhole Deluxe."

Our family did some camping. We had an Army pup tent, no need to dig any foxholes. You taught me how to put it up even though I often had trouble getting the two main stakes to stay up right. All of us enjoyed the camping experience whether it was in the yard or at some park. At our home in Carlisle we did not have much of a yard so Doug and Juddy pretended their room was a tent and decided to build a campfire in their closet. Doug was only about 6 years old at the time but could get a fire to start better than most grownups. To his surprise the clothes that were hanging above his campfire erupted into a conflagration that quickly spread to the attic. Mother was at home and got the fire department out. You were at the Army War College studying hard. When I got home from school that day I was greeted to the spectacle of the fire department and gawking neighbors. Luckily the fire was quickly put out and repairs were minimal. You were not happy about all of this, especially since the landlord would not refund your deposit when we moved to your next assignment. Doug's impressive reputation for "Best Fire Starter" did not end there. Later he would set an entire horse pasture on fire causing much havoc and panic. Luckily none of his fire incidents resulted in any dire consequences, only a lot of hullaballoo and progressively exaggerated story telling.

Another memorable camping experience was our trip to take me to Camp Lachenwald when we were in Germany. I was headed to a Girl Scout Camp and we were all in our station wagon that had the push button transmission, the newest gadget out. We had to travel on some narrow winding roads through some small German towns and villages. I was dozing in the back seat when all of a sudden I heard you and Mother gasp! We were stopped just one foot from a storefront plate glass window full of shoes. The power steering column had broken, leaving you no way to steer the car. It was a new car, a Plymouth we had purchased while we were in Ft Leavenworth, the assignment just before we went to Germany. If this had happened when we were on the Autobahn we would have all been killed!

Love,
Your girl who can put up a pup tent and survive family calamities

Germany—Dec.12, 1944

Dear Mother, Dad & kids,

Not anything much to write but I had a few minutes so I thought I'd drop you a line.

I got a swell package from Robert Wheeler. Is that Uncle Taylor? If so send his address so I can thank him.

This sure is beginning to look like a long war—we seem to be getting weaker and the Germans stronger, I sure wish they would give up.

There has been a lot of "Trench Foot" around. I sure hope I don't get it. I haven't had my socks off for a month, also some of the men have lice. It's just impossible to get clean or change clothing.

I'm still trying to get around to packing some Kraut helmets, insignia etc. and sending them home.

I got your cable—it took almost 3 weeks to get here, it's quicker to write. Don't worry if you don't hear from me, sometimes I just don't have time or facilities to write.

Did you ever get that $140 I allotted to you. You should get a check from the government. Please write soon. Hang on to that 7th Corps Xmas Card for a souvenir.

Love, Jud

Belgium—December 30, 1944

Dear Mother, Dad & kids,

It's been sometime since I've written but there just hasn't been any time. I am now back in Belgium as I have been ever since things started getting hot here. It sure seems funny to be fighting back here again. I've been in some of the same places I was in last Sept. I don't mind freeing this damn country once, but twice is too much.

Spent an enjoyable Xmas Eve getting kicked out of a town by a bunch of Kraut tanks. Those Krauts have the best tank in the world. They captured a lot of our equipment and are using it and dressing their men in our uniforms so you can't tell who is who. I told you those Krauts weren't finished yet. It's nice to be back where the civilians don't shoot you in the back but we aren't getting any closer to Berlin like this. I did get one break on Xmas day, we managed to get mail and I got a package from you—the one with chicken, salad dressing, razor blades etc. It really was swell, thanks a lot. I got another package earlier with 3 cans of nuts which tasted plenty good.

The weather is awfully cold and it's almost impossible to sleep outside. It's even been too cold to snow much. I've never seen it so cold before.

This fighting is a lot more fun than in Germany—the houses aren't torn up and we can sleep inside.

After I left Monchau we went to the Hurtgen Forest, Gey, and Duren—I'm sure glad to be away from there. That forest sure was a gloomy place and the mud was 6 ft. deep (well almost). Here the ground is frozen hard.

Well that's about all for now, I'll try to write soon but our tanks are pretty busy right now. Happy New Year.

Love, Jud

May 19, 2014

Dear Dad,

Trench foot, mud and lice do not sound like fun. I guess you used up your Evening in Paris perfume?

Whenever I think of mud I recall Mother's sister, Kathleen's (aka Auntie Kay, who happened to be my namesake), memorial service and the scattering of her ashes. Remember what a wicked tongue she had? Anyone in her presence felt belittled and embarrassed. Evidently she had to have the last word even as she was six feet under. After Auntie Kay passed away it was mandated in her last wishes that her close family members attend a solemn memorial in Maine, followed by lunch and cocktails and end the day scattering her ashes along the coast. Mother was 83 years old at the time so I offered to take her up to Maine to follow Auntie Kay's final orders. Everything was going well until I noticed that Mother decided that one glass of wine was not enough as she eyeballed Shelley's unfinished margarita. Mother would always say she had to have a margarita because Marguerite was her middle name. Mother polished off Shelley's maragarita. We left the restaurant and headed to the shore. Mother was swaying like the waves below. The day was gray, cloudy and windy, just like Auntie Kay's personality. We arrived at the spot specified by Auntie Kay and began to navigate through mud and rocks to the water's edge. The tide was out. We had, quite sensibly, brought alternate shoes for this part of the occasion. As I guided my tipsy octogenarian Mother by the hand, the mud sucked her shoe right off her foot. We teetered back and forth; a fall into the smelly, mucky mud seemed inevitable. All I could think of was a tumble into the mud would give Auntie Kay such delight as she watched from the great beyond. But we persevered, barely escaped the mud and finally arrived at the place she designated for scattering her ashes. Well, guess which way the wind was blowing? Yes, right into our faces, hair and everything. Yuck. Nothing like having a dead person's ashes all over you.

We eventually made it back to my Cousin Pam's and Mother laid down for a nap in which she slept through dinner and breakfast the following day. Shelley and Cousin Pam were too afraid to check on her the next morning so it fell on my shoulders to see if she was still breathing. Holy cow! Thankfully she was fine, just a little wrung out from booze, mud, and ash.

>Love, Your girl who thwarted Auntie Kay's mud bath

May 20, 2014

Dear Dad,
After I got home I came down with a really nasty bug, ended up with pleurisy and in general was pretty pathetic. I know, it was too far from my heart to kill me but it did make me miserable. Poor baby. Whine, whine. Get tough or die, Kathy. Right?

I am so impressed with your dedication to writing a letter home every week. Was this some kind of agreement you had with Mimi and Samps so they would not be so worried? I remember when Doug and Juddy were in boot camp at Paris Island. They wrote me a letter nearly every day! This may have been one of the only times in their lives they ever wrote a letter, aside from thank you notes you and Mother made sure we sent for Christmas presents. They must have been really suffering. One of them, I can't remember which, made a point to beg me not to have any perfume close to the letters when I wrote back. It seems that pungent letters became a source of great amusement by their drill instructor or whoever it was that distributed the mail. I guess bathing in Evening in Paris was out of the question. I will never stop teasing you about your perfume.

Did Doug ever call you for hours on end when he was in Kuwait loading up equipment after the Persian Gulf War? When I asked who was paying for the long distance call he said "some Sheik guy." He would tell me about the women, whom he referred to as "Ninja

Women" due to their affinity for total black. It also seems that the laborers who cut the sparse lawns did so with clippers.

Love, Your poor baby whiney girl

Fred, Dave & Dad-The Bridge Brigade

Hippie Daughter Kathy

Liberty's Wedding-Escorted by Travis and Ross

Loralei, Liberty & Travis

JANUARY 1945

Belgium—Jan 1, 1945

Dear Mother, Dad & kids,

Well I was notified today that I was promoted to First LT., effective Dec. 26, so it was a pretty good New Year's day. The promotion carries about $15 more per month, which is okay with me. I also got your package with the camera in it so tomorrow I'll start taking some interesting pictures, including one of my platoon and a tank.

Enclosed is a surrender ticket that our air force drops to the Krauts to get them to come over and surrender. The package was swell—that was the first tuna fish that I've had in a long time.

I got my head shaved the other day and I've got an awful cold. I've also got a swell mustache, you ought to see me. It's just too much trouble to keep my head clean with a lot of hair. The weather is still very cold but we haven't had much more snow.

We haven't been getting many letters because of the Xmas packages; I hope they start catching up soon. This cold weather is awful, especially when you have to be out in it all the time. I'll sure be glad when spring comes.

That issue of the World kind of tickled me—it said it looked like the Germans are going to retreat back to the Rhine—looks like they were just a little wrong.

The Limeys (English) are sure a screwy bunch. The other day one asked me if I wanted a drink, I said yes thinking he had some cognac—what did he pull out but a thermos bottle of tea. I nearly died laughing. All their equipment is American, jeeps, tanks & everything. We got a new type of sleeping bag today. I got two of them and put one inside the other. I'm going to try it out tonight—it ought to work pretty good.

Would you buy some 1st lieutenant's bars and send them to me please? I can't get any here. About 4 pair would be swell. Write soon.

Love, Jud

May 31, 2014

Dear Dad,

Congratulations on your promotion! I am sure it was welcome news on a frigid New Year's Day still in the throes of the Battle of the Bulge. You definitely earned your rank the hardest way possible in the military.

I am glad that you enjoyed the tuna fish enclosed in your last package. Did you ever hear about the time that Juddy made himself a "tuna fish" sandwich back when we lived in Tampa? Well he did not look at the label very closely and the can of "tuna fish" just happened to be a can of "tuna" flavored cat food; pet food code for mystery meat with "eau de tuna." Well he gobbled the finished product with zest and relish. Just as he was swallowing the last bite, Shelley came in to feed the cat and discovered the can of cat tuna missing from the refrigerator. She confronted Juddy and the two of them figured out what he had eaten. Poor Juddy was sick and crying mad. We were not all that sympathetic and it only increased his fury.

It was also around this time that I made a batch of chili. I would make it from scratch and we all liked it. After all, how can you mess up chili, right? I had all the ingredients cooking well and it smelled very inviting. I only had to add the chili powder. As I sprinkled in the powder something seemed out of whack, there were slight movements in the pot. To my horror I realized that the chili powder was full of those tiny brown bugs. I did not know what species they were or if they were toxic, hallucinogenic (after all this was the 60's) or flavorful. I debated. Should I say nothing and feed it to the kids or should I throw it all away, a total waste of food, not to mention my labors over a hot stove. I seemed to remember from biology class that insects were high in protein. Protein is good for growing kids, right? For some reason I stepped out of the kitchen before deciding what to do. Juddy, however, was one who always made his way to the kitchen as soon as he came in the house. Before I knew it he was plopped down in front of the TV with a bowl full of the insect ridden chili concoction. To my great shame and amusement I said nothing until he had eaten the entire bowl. I studied him closely for any adverse reactions. There were none. In fact he headed to the kitchen for more. Then Shelley and Doug came in and I told them all about the bugs. Now there was a reaction, one of rage! Once again we were not all that sympathetic. Did you ever have to eat food laced with bugs? I am sure if someone were hungry enough a few bugs would not be much of a deterrent. I hope you were never in that situation.

I am relieved to hear that you received adequate sleeping bags. It saves time making a bed as well. Due to your military background, you reasoned that we would all know how to properly make a bed, the corners neat and tidy and all the wrinkles smoothed out. To this day I cannot stand an unmade bed, it drives me nuts. The first thing I do when I get up is to arrange my bed with those neat and tidy corners. I made my kids do the same thing. Travis did figure a way around it for a while, he put his sleeping bag on top of his made up bed and slept in the bag instead of getting under the bed covers.

When we lived in Germany you sanctioned Saturday morning inspections. Every Saturday we had a written list of chores to be completed and crossed off. One Saturday, that darn Shelley, with whom I shared a room, had hidden some food between the box springs and the mattress, squirreling away food for winter. When you inspected our room of course you found it. I was the oldest and therefore responsible. I was mad for a while but I learned to check that mattress for her food stash. Today she is "Miss Nanny, Nanny, Neat, Neat." She makes you take your shoes off and everything. I am not sure what caused this transformation but if one could put it in a bottle and market it you would become rich!

Love, Your girl with evil cooking distinction

P.S. That is pretty funny about the offer of a cup of tea!! One lump of sugar or two?

Belgium—Jan 3, 1945

Dear Mother, Dad & kids,

Today at mail call I got a package (the one with the pictures) and two letters. The package was really swell, especially the photos—thanks a million.

That Thompson you asked about was Kenneth—I'm sorry to say he was killed shortly after he joined us. These new guys just don't seem to last very long—they just don't know the tricks of keeping alive.

This watch I've got is lousy—I'm trying to get a Kraut with a better one.

There is one thing I wish you would send me and that is a pair of fur lined leather gloves or mittens. Mittens are preferable but they must have a trigger finger. I have an awful time keeping my hands warm in these army gloves. Also I'd like an Olive Drab muffler.

I'm permitted to name a couple of places in Belgium near which I fought recently—Marche and Rochefort, you probably read about them in the papers.

Some of these Belgians sure make me mad. The other night I went up to a big house and asked if I could use a couple of rooms. The man spoke English and no, he didn't want a bunch of soldiers dirtying up his house and we would have to sleep outside (it was below freezing and snowing like everything. I told him to go to hell and let him sleep in the barn and we took the whole house. If he had been decent we would have just used a couple of rooms. We can certainly expect that much from these people. Most of them are pretty nice though.

Well that's about all for now; say you might put a couple of those 1st lieutenant's bars in an envelope, they would probably get through.

Love to all, Jud

June 9, 2014

Dear Dad,

I am sorry that it has been so long since I last wrote. Shelley, Doug, and I have been in Tulsa. We went, in part, for me to do some research on the book and also to try and visit some of the Tulsa Millers.

We began our visit with a trip to the now nefarious bridge over the Arkansas River where, as Fred stated, "Your Dad tried to kill us." Today it is very picturesque and part of a hiking, running, and biking path along the river. I tried to drag Shelley across in an enactment of your journey years ago but she would have none of it! Ha!!

The next day we went to OMA for the Reunion Opening Reception. We went early to take a tour of the museum and see your photo in the OMA Hall of Fame.

At dinner we met Jack Harris, class of 1955. He said that he was a Company Commander for you in Bad Kissingen in the 15th Armored Calvary when we all lived there after the war. He remembered our mother's name Bette and a wacky yet memorable story about a pony. Did you bring a pony to the Officer's Club and hand out mugs full of "pony piss"? Jack said that to this day he still has his mug. I wonder if these mugs were like the naughty beer stein we used to have. I am referring to the one that when the bottom was held up to the light a naked woman was revealed. I didn't feel comfortable enough to ask Jack about the details of his mug. How were you able to get the pony to produce enough "liquid" for the mugs? Was it really from a pony or another species? What were you thinking?

Jack also shared a moment when you looked over his tanks and remarked they were not up to par and he was visibly upset. He said you knew he was disturbed but in the final analysis his tanks earned the highest evaluation. He seemed to hold you in high regard, saying that you had to earn things the hard way by coming up though the ranks.

We also toured some of the grounds and even though OMA is no longer in existence, they have done a good job of preserving its history, for example replicating your quarters.

I also found an old newspaper clipping of you and fellow soldiers polishing your boots!

We spent most of Saturday at the reunion attending a luncheon followed by the debut of a made for TV documentary about OMA. I must confess I was disappointed there was not much time devoted to recognizing those who became Generals from OMA. I was somewhat puzzled because so many who did not make a career out of the military had prominence in this film. This seemed somehow to diminish it in my eyes. I wonder what your opinion would have been about the documentary.

We also toured the grounds wearing our "Dear Dad" shirts that we had made for the occasion. The shirts have a photo of you as a private that dramatically contrasts with the one of you as a General.

You look extremely young in the first picture. What a baby face!! You always looked much younger than your years, except in the ones where you have been in close battle. We received a lot of compliments on the shirts as more and more children and grandchildren are attending these reunions.

In an attempt to connect with anyone else that might have known you or remembered you, I was given permission to set up a display asking for information about you, but to no avail. We also tried to visit with the Tulsa Millers while we were there but Aunt Betty was too sick and not up to any company. That left us with most of Sunday to look around the old neighborhood and catch a movie. Shelley and I wanted to see if we could find the deli where Uncle Fred had enjoyed some notoriety in his elder years.

Do you remember an eatery in Tulsa called Lambrusco's Deli? Uncle Fred used to love going there, especially after his wife died. His favorite item was the chicken Caesar salad he would generously share with his dachshund named Ziggy. One time when Shelley and I were visiting, Uncle Fred treated us to lunch from Lambrusco's. As we started to leave we asked Uncle Fred if he wanted to let one of us drive but he said no, he would drive. The problem was that Uncle Fred no longer held a valid driver's license and his tag was expired. When we asked him about this he just shrugged it off and said that you could "Spring" him if he got locked up (by now you were a practicing attorney in Washington state, not much help to him in Oklahoma). Well off we go, totally illegal, cruising the streets of Tulsa in search of nourishment. He took some back streets in an attempt to elude the authorities he thought might be lurking nearby in search of elderly law breaking curmudgeons. We arrived at Lambrusco's without any guns drawn or "keep calm and back away from the curmudgeon." Shelley and I started to get out and he said no, they would come to us. Hm, curbside service? To our astonishment, a well-endowed, beautiful girl of about 20 something appeared next to our car. She had impressive cleavage with an intriguing tattoo between her breasts. It was impossible not to stare. She said to my uncle of 80 plus years, "How ya doing, Studly?" Ha!!! But it got even better. Before we could recover from hearing her call him "Studly" she plants a big juicy kiss dead on the lips of my Uncle, the "Stud." Uncle Fred turned to me with what was the hugest "shit eating grin" I ever saw in my life. No wonder he drove illegally to get to this place!

Thankfully we found Lambrusco's Deli on our last day in Tulsa but it was closed that day. The old neighborhood is still as beautiful as before. I did get a directory from the reunion that I am going to follow up on and see if there are more stories about you out there.

Love, Your girl who loved her Uncle "Studly"

Belgium—Jan 9, 1945

Dear Mother, Dad & kids,

I'm so cold I can hardly hold this pen. The snow is terrific. I never saw such cold weather or so much snow. The Krauts sure know more about winter fighting than we do. They have white uniforms and paint their tanks white. Also their uniforms are much warmer than ours. We caught a Kraut with some shoes on that he had taken off a dead Yank so we made him walk barefoot back to the PW enclosure;

I guess he froze his toes. Mine nearly freeze even with 3 pair of socks, shoes, and overshoes on.

I got your letter, David, those articles were really okay. Get in the navy; they have it a lot better than we do. No mud or cold.

If you send any more packages here are some things I would like to have. A toilet kit in a leather case (as small as possible and with a metal mirror, razor, blades, soap box etc.) and a box of cocoa. Also those mittens and 1st Lt's bars.

Tomorrow if I can get some money orders I'll send some more money home. Just how much do I have now? Please write soon.

Love, Jud

June 14, 2014

Dear Dad,

It is harrowing to read about how cold you were and your constant plea for warm clothes sent from home. Trying to stay alive and fight is stressful enough without having to suffer extreme temperatures and not have sufficient warmth and protection. The closest thing I can remember to toughing it out in the cold were a few winters while training race horses for a living. We did not have enough money for propane gas, just got up and had a hot shower then headed to the barn to work the horses. As long as we stayed busy and moved around it was bearable but nighttime was miserable shivering in front of the television. This, however, was a Georgia winter and did not come close to the temperatures you endured. I read that the average temperature in Belgium during the winter of 1945 was -4 degrees Fahrenheit. The ground was frozen solid and there were relentless snowstorms.

We did have one really cold Christmas during my horse training days. The kids and I were in Valdosta, Georgia, and a winter storm was approaching. This time we did have enough money for propane gas and I was keeping an eye on the storm daily as well as monitoring my propane tank. I called the gas company one week prior to Christmas for a delivery, knowing that at the rate I was using gas we would not have enough to last through Christmas with the impending storm. As each day passed the storm drew nearer and the gas did not arrive. This storm was one for the record books with predictions of snow for South Georgia and North Florida, something that is extremely rare. On Friday the storm hit and the gas company still had not made their promised delivery, stating that demand was extremely high as residents prepared for the storm. I had been calling for a week. By 10 pm Friday night the snow was deep enough that the kids made a snowman and were throwing snowballs as they enjoyed this rare event. I had made chili (no I did not have bugs in it, ha, ha). The heater was running nonstop, down to less than 1% on the gas gauge. That night the temperature fell to 18 degrees, unheard of in that part of Georgia. The high temperature the next day was 22 degrees. The heater continued to run nonstop. The next morning we were faced with no water at the barn for the horses. We had to take a 50 gallon drum, drive the truck to the bottom of the hill where there was a pond, dip buckets of water from the pond into the drum, drive back up the hill and repeat this process until we had watered 33 head of horses. It took us 2 ½ hours just to water them, something that normally took about 20 minutes. Back at the house our gas ran out around 3pm. We turned on some of the taps to keep the water going at the house and made a trip with a small 30-gallon propane tank to find gas. This was no small feat as there had been ice accompanying the snow and the highway patrol had closed I-75 and prohibited gas trucks from driving. Most of the roads were impassable due to ice and downed limbs. We did get the heater going again but the pipes froze so we had no water at the house. I had a sink full of dirty chili dishes and no way to wash them. The low that night was back in the teens once again. The following day was Christmas Eve. We were excited because it was going to reach 34 degrees that day! The water for the horses finally defrosted but at the house it still was not running. I took the kids to Wally World for some last minute shopping. When we returned home the water had decided to flow once more and was running straight down the hallway from the bathroom into the living room, soaking the carpet, under the Christmas tree, just everywhere. The bathtub tap that had been left on unfroze causing the tub to fill up and overflow.

I took out a space heater we had and placed it in the living room to assist with drying the carpet. Then I retrieved my hair blower and went from one end of the house to the other furiously determined to get my home back in some kind of civilized order. I began around 4 pm and finally had the dirty dishes clean, the carpet dried, and the Santa chores completed by 1am. What an unforgettable Christmas.

> Love, Your girl who loves running water and heat

Belgium—Jan 11, 1945

Dear Mother, Dad & kids,

I got a letter from you today dated Dec. 19, the one with Capt. Hamilton's letter.

Today the weather was very cold but clear and I took some good pictures. I'm having them developed here as the censor has to pass them. I took some pictures of my platoon, some Kraut tanks, dead Krauts, and also some pictures of some of the American equipment (Tanks etc.) that the Krauts used in their attack. Most of the stuff they used was American, they must have really captured a lot of it.

My fingers are so numb I can hardly write, I sure hope this snow melts soon but the Belgians say it stays till late March.

The civilians sure don't give us the welcome they used to. I'm getting so I dislike all foreigners.

It looks like the Krauts are pulling back into Germany but I don't doubt that they will attack again as most of the equipment they are losing is stuff they captured from us. I wouldn't mind this war half so much if it would only warm up a little. I never saw anything like it.

Thanks a lot for getting the perfume for Patti, I tried to get something for everyone while I was in Verviers but I couldn't find a thing. Please write often.

Love, Jud

June 21, 2014

Dear Dad,

Your firsthand account of The Battle of the Bulge is fascinating. I am so proud to be your daughter.

I hate that you were still so cold. Travis and I once ran a race in Macon, Georgia during the middle of a winter ice storm. There were two races that day. Travis ran the 5K while I ran the 15K. I was warm enough while I ran but something weird happened to my eyes. My eyelashes transformed into little icicles! When I got back to the van Travis had already warmed it up and removed his socks and wet clothes, trying to dry them out on the air vents. He complained about my lack of sympathy for his suffering. After all he only suffered for 3.1 miles and he was very fast so his pain was short lived while I had to endure 9.1 miles of that stuff. I guess I was a terrible mother. I believe I even told him "Poor Baby." He won his age group and I won mine but to this day he threatens to turn me into the authorities for child abuse, stating I shamed him into running that day. Ha!

I must share that I am worried about my cat. I know that you could care less about the four-legged creature, but there is something very wrong with her. Yes, I know she is 16 years old; she has lost a lot of weight and can't keep her food down. She is eating so that is good but she is losing a lot of hair and that is not good. It seems like death is going to visit me again. So if you are channeling through the cat you might need to look elsewhere. In the last 7 years that I have been in this swamp place called Waycross, Georgia, I have buried my two horses, my dog, and lost you and Mother. I will be glad to retire and get away from this place that seems to only be filled with memories of death and dying. I am not comfortable living here. I sometimes feel like you did, always living in a foreign land. I feel so out of place, like living in the Twilight Zone. People here don't watch the news; they get it from their preacher. If you are white you are supposed to hate blacks and Hispanics and believe that Obama was born in Kenya. They look at me like I am an idiot when they learn that I am studying Spanish and International Studies. I guess this is a good place to get a taste of what must be culture shock. I should be well prepared for a stint in the Peace Corps; at least that is what I will tell them.

Boy I sure am whining a lot tonight. Poor Baby. Maybe I should go seek refuge in a carton of chocolate chip mint ice cream.

Love, Your girl who sometimes whines

Belgium—Jan 15, 1945

Dear Mother, Dad & kids,
Nothing to write as usual except about the war and the weather. The weather is very cold and still lots of snow. The war is going pretty good and I guess before long I'll be back in Germany.

I got a couple of good snapshots today. One of one of our bombers going down in flames and another of a 16 year old Kraut we caught—he isn't as big as Fred.

I hope people at home don't believe everything they read in the papers, about 50% of the news is a bunch of lies—I know because I'm here where the news is made.

Say, on some magazines they have special small size overseas editions, how about seeing if you could get me subscriptions to Life, Time, Esquire, Cosmopolitan, Newsweek, and Liberty. Whenever we do get a break about all there is to do is sit. Please write soon.

Love, Jud

June 22, 2014

Dear Dad,

I was not aware of special small editions of magazines; you always were an avid reader. I don't know how you juggled reading at least four books at the same time, unlike me preferring to read one at a time. One of our favorite places to go was Barnes and Noble. We would browse, buy some books, and then top the time off with a Starbucks. We were always exchanging books. One of the last that I gave you was Tom Friedman's *The World is Flat*. I used to email you his column and when I would call you at night we would discuss what was going on in the world. I really miss that.

I know you enjoyed following the news and reading magazines. Do you remember watching the Huntley Brinkley report with me? Were you skeptical that news wasn't being portrayed accurately? Fox News is nothing more than a propaganda machine similar to Goebbles in World War II.

We also loved to watch Forensic Files (nothing like a good murder right before bed) and Jeopardy. You were hard to beat on Jeopardy but every now and then I could out guess you.

You also brought back some interesting artifacts from your travels. One of those artifacts was a set of African ebony heads; they were beautiful. At least they were beautiful until the day Doug and Juddy got into a fight and Doug threw one of the heads at Juddy. It missed Juddy but hit a wall or something, breaking a piece off the base. We carefully glued it back on but this dastardly deed did not escape your eagle eyes. There were stern words from you and downcast eyes from Doug and Juddy. You will be amused to know that Shelley and I unanimously decided the African ebony heads were to go to Doug to preserve his fond memories of his pugilistic skills. Mother was the one who had the best consequence when the two of them resorted to violence instead of peaceful diplomacy (Ha, ha yeah right). If the fight erupted outside, she would turn the hose on them and "cool" them off, very effective. In spite of childhood battles they became best friends and went into the Marines together. You were so proud of them! We all were. One thing I always wondered about. Were the Marines able to teach them to get the ring in the bathtub clean? When they took a bath at night they were supposed to clean the tub but they always did a terrible job. It usually looked like a dog had just bathed instead of two boys. When it was their week to do the dishes they never really got the ick off the pots and pans either.

I do need to confess about a time that Doug took the rap for drinking your coke. I was the culprit but he confessed so he could get the licking over with and go outside and play. I never said a word. I was a bad sister that time. He still reminds me of my crime today but seems to have forgiven me. Or maybe not since he mentions it every time we get together.

I love looking at photos of them. Sometimes they look like little angels that could have been a part of "Spanky and Our Gang" instead of "African Head Throwing Banjees." Doug has a photo that reminds me of Alfred E Newman on Mad magazine!

Love, Your girl who always loved her brothers

Belgium—Jan 19, 1945

Dear Mother, Dad & kids,

Enclosed are money orders for $100 to be put into bonds. On that check you got it should have been $140.

In the Stars and Stripes (our GI newspaper) there was a big article on our outfit capturing Laroche. I was going to send it to you but someone else cut it out. Anytime you read about the 4th Calvary, that's me. You probably read about Laroche in your paper. It sure was a mess, our Air Corps had bombed it flat. In one air raid shelter 200 civilians were trapped and died.

The other day we were taking it easy and since I had the day off I took a jeep and went to Bastogne—it's not torn up as much as I thought it would be.

I sent 2 rolls of film back to be developed. They should be ready in a week or so.

It snows about every other day and never melts so we have quite a bit on the ground. I guess it should start getting warm in about a month.

Our mail hasn't been coming through so good lately; I don't know what the trouble is.

I'm pretty sleepy so I'll close for now. Please let me know just as soon as you get these money orders and what my total is.

Love, Jud

June 23, 2014

Dear Dad,

I took the cat to the vet today. The preliminary tests show an extremely high white cell blood count and an extremely low red blood cell count. It does not look good and I am depressed so I decided to look through some more photos. Aha! Guess what I discovered? You are holding a cat with what appears to be tender loving care! Boy what some guys will do to get a dame. This was June, your second wife's cat.

I am not buying it Dad. Not one bit. I know how you really feel about cats. After all, you and June did not keep any cats for long. This is further proof of this obvious ploy to win her affections. Look at where your hand was, clasping at the throat. I bet you were thinking, "just a quick snap of the neck." That is what that smile is really about. Run for your life, cat!!

Love, Your girl who knows the real reason for your smile

Shelley-Miss Nanny, Nanny, Neat, Neat

Trying to drag Shelley across Dad's childhood bridge of terror

OMA Hall of Fame

OMA quarters

OMA bunks

Dad polishing boots

In front of George Washington statue at OMA

Dear Dad shirts

Uncle Studly

Did You Know My Dad? Judson Miller Class of '42 Please tell me about him!! Thank you!

HALL OF FAME
1988

BRIG. GEN. (RET) JUDSON MILLER
CLASS OF 1942

Miller entered OMA as a 10th grader in high school following graduation from Woodrow Wilson Junior High in Tulsa. Two days after his 17th birthday, which was December 5, 1941, Pearl Harbor occurred and set in motion events that would govern his life for the next 33 years. In December 1942, Jud and a small group of cadets enlisted in the army for service in the war.

After being selected for Officer Candidate School Miller was commissioned as a second lieutenant in 1943 and served in the European Theater as an armored cavalry officer. It was in Europe that he earned the Purple Heart for wounds suffered in combat.

He also saw combat duty in Korea, where he became one of the youngest majors in the United States Army, and in Vietnam where, as a colonel, he directed the preparation of a 4,200 man brigade of the Fourth Infantry Division for combat duty and led the brigade in combat against the communist forces.

In addition to the Purple Heart Miller has been decorated with some of the nation's highest medals and commendations for both meritorious achievement and valor in combat, including the Silver Star and Bronze Star, Joint Service Commendation Medal, and the Distinguished Service Medal of Chung Moo and the Presidential Unit Citation, both from the Republic of Korea. He was also promoted to brigadier general at the age of 43.

Miller retired from his military service in July of 1976, and retired as a brigadier general. At age , Miller enrolled in law school, obtained a law degree and passed the Washington State Bar and currently practices law in Tacoma, Washington where he now resides.

No longer young

Doug in his Alfred E. Neuman look

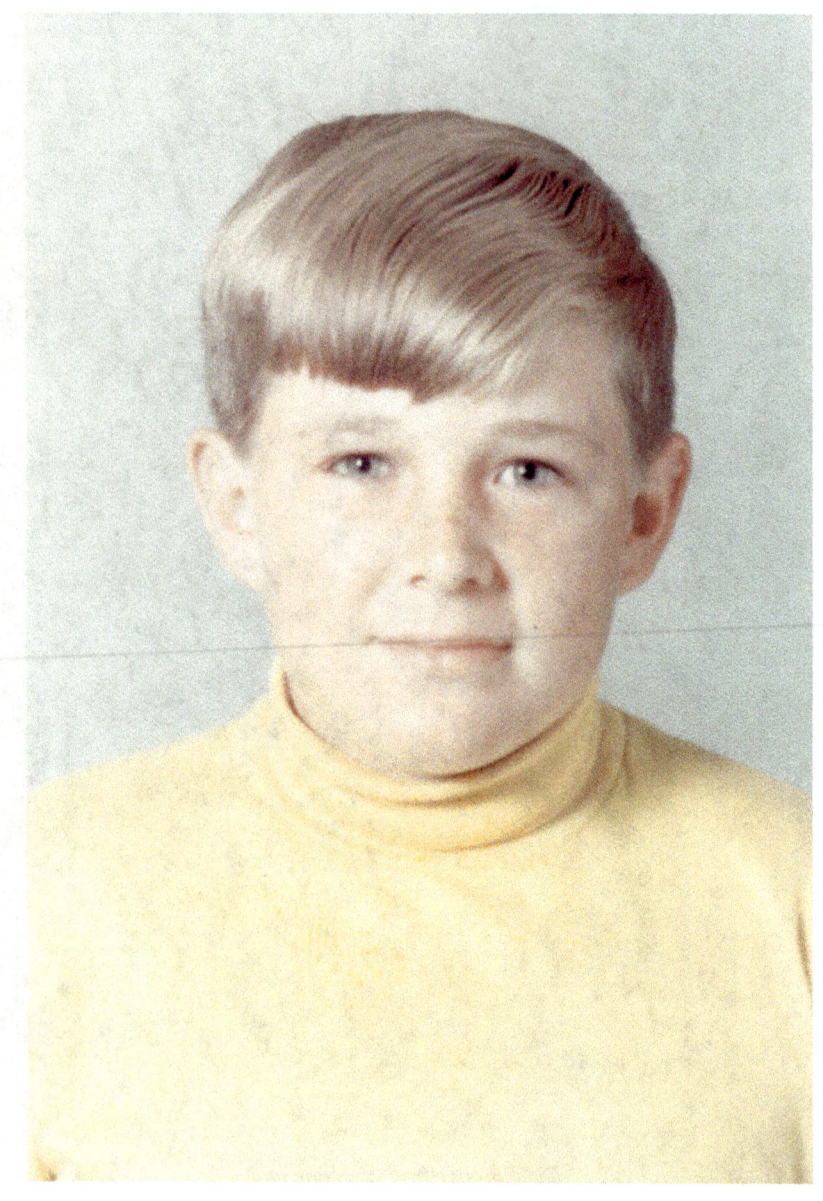

Juddy in his Spanky & Our Gang look

Doug as a Marine

Juddy as a Marine

Run for your life, cat!

FEBRUARY 1945

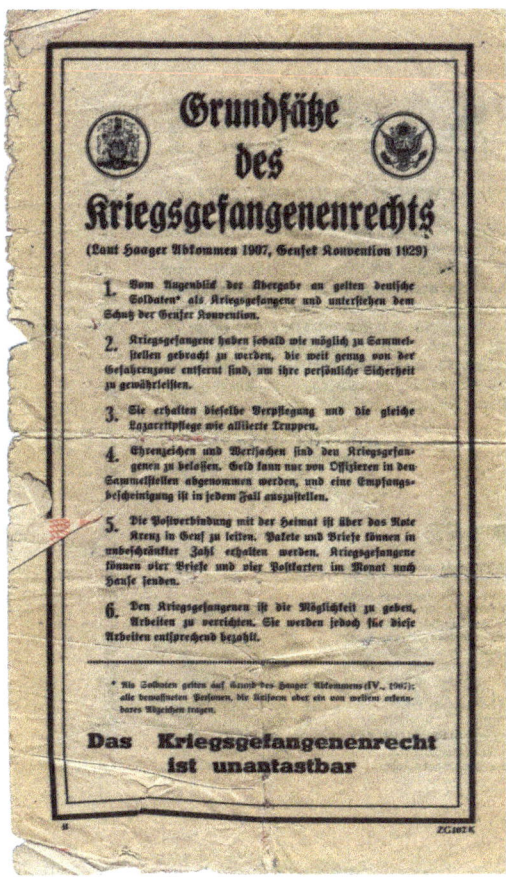

Belgium—Feb 1, 1945

Dear Mother, Dad & kids,

    I've been pretty busy lately and haven't had time to write. I just got back from spending a few days with the Air Corps at a P-58 base. I was one of a group of front line officers sent back to tell the Air Corps what it is like up here and see how they live and work. They sure have it easy. Beds, electricity, running water, big cities with nightclubs and everything. Also their casualties are *very* light. Enclosed are 4 photos I took with some borrowed film—not much good. I've got a large supply of film now that I got while I was back with the Air Corps—they have all the comforts and luxuries and we can't even get soap, it just isn't fair. Notes on back of film tell what they are. I have 2 rolls back being developed now and they should be much better. I wish I had gotten some better pictures of Laroche—I never saw a town so beat up. I've gotten several packages lately; the one with the watchstraps and the sweater—everything was fine and very practical. Thanks a lot. I'll try to write a longer letter tomorrow.

                                                      Love to all, Jud

June 25, 2014

Dear Dad,

    I am sorry that you were exposed to life in the Air Corps. It is almost better to be oblivious to what is around you rather than see reality. How can one soldier have a bed and another not even have soap? Why was there such disparity between our American troops? There is little doubt that the soldier on the ground bears the brunt of the hardships of war and I do not think that will ever change, no matter how technology revolutionizes combat.

    I wish I had stayed oblivious about the condition of my cat, but reality has smacked me in the face once again. Gracie has cancer. The vet prescribed medication to help with vomiting but now I have to decide what to do. Do I just try to keep her comfortable and let her die on her own or do I have her put to sleep to save her from suffering? If I put her to sleep whose suffering am I really trying to alleviate, mine or hers? Am I being selfish or am I being compassionate? As irrational as it sounds, I have this idiotic desire to ask the cat if she has a living will. That would solve my moral dilemma, my angst, my indecision, and my guilt. "Hey Gracie, do you have a living will? Is it in your safe deposit box? Who is your lawyer? Who gets your catnip mouse when you die?" She would probably just tell me not to worry she still has 8 lives after this one. I realize this would not cause you great anxiety but if you are residing in the cat you might need to start worrying. Obviously I have read too many Stephen King novels to be entertaining this line of thought.

I lie awake some nights and think about how you suffered towards the end of your life. It was horrible when you were unable to sleep and you would pace, saying that you wanted to kill yourself if you could not get some sleep. You had a book, *Final Exit,* which terrified me but made some kind of awful sense. I used to look for your .45 that you kept under the bed and was both relieved and horrified when it went missing. What a dilemma death can be.

                Love, Your girl who must come to terms with reality

Germany—Feb 4, 1945

Dear Mother, Dad & kids,

Well here I am back in the fatherland and none too happy about being here.

The weather has warmed up considerably and the snow is almost gone but we expect it back again.

One thing that makes me laugh here in Germany are all the Esso and Shell Oil Co. signs everywhere.

Did I tell you I had received several packages—the one with the sweater and the one with the watchbands. Thanks a million. One thing I would like to have is a good durable billfold, the one I have is kaput.

The war news looks good but I'm afraid the Krauts will keep on fighting for a long time.

The newspaper is coming through fine—please see what you can do about some magazine subscriptions. Please write often.

Love to all, Jud

July 2, 2014-Liberty's birthday

Dear Dad,

So it is back to "The Fatherland," land of lederhosen, beer, and a few Nazis here and there. Quit your whining and go use your hunter-gatherer skills to locate a brew or two. I know you can do it.

I have been at a convention for work on Amelia Island, Florida. It is the last one I will be attending. Liberty went as my date. Over the years I have taken Shelley, Doug, and all my children. One of the events is off-shore fishing. Liberty and I did that one this year.

She and I also went to Cumberland Island and spent the day hiking the ruins of Dungeness as well as the pristine beaches. Dungeness is such an incredible sight even as a ruin. What it must have looked like before its decay and fire! The horses seem to enjoy its lawn now.

We discussed Gracie as I value her opinion as a vet and a loving daughter. She had the same crisis with her old dog Swain, sharing with me her regret for not putting him to sleep sooner. So some of my cat angst has abated, at least for today. Just exactly what is the right thing to do changes with me, moment to moment. I realize that if you were here you would have no problem dispatching the cat up to heaven or wherever it is that they go. Maybe they go to the land of milk and mice. Hmm. Gracie would prefer a place with

"Fancy Feast," especially the "Elegant Medleys." She also likes a piece of "Sargento's Natural Baby Swiss" cheese as a dessert from time to time. She would also need some birds for her amusement. When she looks at birds she foams at the mouth and starts smacking her cat lips in anticipation of "Raw Bird." Maybe she refers to them as "Bird Tartar." Yum, yum delicious.

Love, Your girl who wonders about cat heaven

Germany—Feb 12, 1945

Dear Mother, Dad & kids,

I got Fred's letter today, also one from Grant, one from Doris Mullins, and two from Patti, so my morale is really high.

Weather is terrible—it rains during the day and snows and freezes at night. However I'm in the best of health despite the weather and the Krauts—hope I keep that way.

Did you ever get the $100 I sent home?

I got a new shoulder holster for my .45 pistol today—its a heck of a lot handier and I can strap more ammunition for my tommy gun on my belt.

These German civilians sure make me mad. No fooling if I had my way we would shoot all of them, then there wouldn't be any more wars.

Guess what—I finally sent a box of souvenirs home. Included were: a Nazi flag, a Nazi armband (taken off a dead SS officer), some shoulder straps I got off dead Kraut officers, an iron bar that SS troopers held in their hands to beat prisoners, and some other junk.

Answers to your questions—1. Major Donnell is not one of my officers, 2. I am well 3. Yes, I got the package with the photos (they were swell and I would like some more pictures) 4. I get all of your letters so there is no need to send duplicates.

My candle (taken from a German home) is about burned out so I had better close. Please write soon and send some more photos—also a new billfold please.

Love to all, Jud

July 4, 2014

Dear Dad,

Happy 4th of July. Today I ran a 5k on St Simon's Island. My time really stunk and I ended up 4th in my age group. It was hot and humid, which did not help my depression about my cat. All Gracie does now is eat, vomit several times a day, and then meow to go outside. She now prefers to sleep on the front or back porch, not wanting to stay inside anymore. I wonder if she is cold but I have no real way of knowing. She is losing more weight but does still like for me to pet her.

Love, Your girl depressed about a cat

Germany—Feb 28, 1945

Dear Mother, Dad & kids,

As you have probably seen in the papers things are moving pretty fast and I haven't had much time to write, besides I can't ever think of anything to write—we can't tell what we are doing because of the censor.

I got the package with the bars- thanks a lot. I don't usually wear them on my outer coat because the Krauts always shoot at officers first.

The weather has been pretty warm lately but plenty of rain.

One thing I would like to have is some candles (lots of them) and a good leather billfold.

I sure wish I would get a pass to Paris—other outfits get lots of passes but our colonel hates to see anyone have a good time. He is a really hated man—I wouldn't be surprised to see one of the men shoot him someday. He never is up where the fighting is—he makes me sick. I sure have a good captain though—the best I ever had. I'm now the highest ranking officer in the company besides the captain, the other 1st Lt got killed the other day.

David wanted to know what division I'm in. I'm not in any division, we are just the 4th Calvary Regiment or Group as they call it. We never get any credit for what we do. Like when we captured Laroche the Tulsa paper said it was taken by the 84th division—baloney!

Please write often and send me some more food—everything you sent was swell and arrived in good shape.

Love to all, Jud

PS. Please, Please, Please no more V-mail—it reminds me of a dried up letter.

July 5, 2014

Dear Dad,

I am not going to write about the cat. It is too depressing and seems pretty stupid to suffer over a cat when I read about what you endured. You have probably seen many of your best friends killed so a dying cat seems pretty insignificant.

I looked at the map and all I can tell, with my ignorance of Army map reading, is that you guys were headed to the Rhine at this time. I wonder if you were anywhere close to the Lorelei, a large rock on the Rhine River that means, "murmuring rock." A German author wrote a poem about a beautiful woman who would sit on this rock combing her hair and tempting the sailors who passed by. Because of her distracting beauty, sailors would crash into the rock and meet an early death. It was said that the rock echoed her longing for a lost love. I was so captivated by this legend as a child that I named one of my daughters Loralei. I guess you were not all that interested in the tale of the Loralei since you had Nazis singing death threats with their guns.

It is said that old people think you "mark" someone with the name you bestow upon them. I thought about two confrontations that she and I had. The first one was when she was around 3 or 4 and I spanked her for some infraction. She told me, "That didn't hurt and I hate you." The next day we were coming back from working the horses at the track and she started provoking again with her smart mouth. Leroy, her step-dad, pulled the truck over, got out, and broke a switch from a tree. He looked at her and without a word placed the switch on the dash of the truck. We arrived home and she started again so he (for the first time) took the switch and switched her legs. She came running to me and said, "Look what he did!" and pointed to a small red mark. The next day she took her coloring markers to the track and decorated a switch of her own. She brought the "switch of many colors" home, pointed and daringly said to Leroy's face, "This is for you when you are bad." She then placed it alongside the switch that Leroy had used on her the day before. Thankfully she transformed into a very nice sweet little girl!

The second confrontation was in her senior year of high school, just prior to her graduation. We got into an argument in church and I got so mad I grabbed her and Travis and we left. She kept on with her smart mouth, pushing my buttons saying, "I know you want to hit me." Finally I had enough and I pulled the van over, got out, jerked her door open, and pulled her out on the side of the road. I told her, "Okay, you are 18 and you want a fight—come on hit me." Travis was crying in the backseat, "No Mama, no." She would not hit me. I really wanted her to. Weeks later she left for college telling me, "I hate living with you." She stayed away three weeks. Then all of a sudden she started coming home every weekend telling Travis, "You really don't know how hard it is to live on your own." My Loralei is definitely a temptress as far as I am concerned; tempted me to commit bodily harm that is. Today she is one of my greatest joys!

I did dig out a newspaper article about the capture of Cologne where your outfit participated in raising the flag. Cologne fell on March 7, 1945. No wonder you can't say where you are. You are poised to cross the Rhine!! Crossing the Rhine is a very big deal!!

I am sorry you had a lousy colonel; I too have a crappy boss. My boss calls himself, "Tiller the Hun" which is laughable since he has man boobs and the most athletic thing he does is play golf. You and I have discussed him before. "Tiller the Hun" (this was how he introduced himself to us) uses termination (not the final kind, just you no longer have a job kind, no way to feed your kids etc. kind) and humiliation as his weapons of choice. He lacks any means to inspire or bring out the best in people. The only time he is sincere is when he humiliates someone. When he offers praise it lacks all credibility and sincerity. He smiles with his mouth and hates with his eyes. What a phony. If your colonel is anything like "Tiller the Hun" someone will shoot him before the war is over.

It is unfortunate that you were having so much trouble getting the folks back home to get you a new billfold. Seems funny how little things can be so annoying.

    Love, Your girl who would send a billfold if she could

Group of refugees Laroche, Belgium

Part of Laroche nearly destroyed

Liberty lands a bonnethead shark

Cumberland Island horses and ruins of Dungeness

Gracie napping

Old Glory Hoisted By VII Corps in Cologne Ceremony

By John Thompson
(Chicago Tribune Correspondent)

COLOGNE.—The American flag was raised recently on the Rhine for the first time since 1921 when Old Glory was hauled down at Ehrenbreitstein as the last infantry regiment in the army of occupation pulled stakes and started home.

The flag isn't fluttering at Ehrenbreitstein now for that town across the river from Cologne is still held by the Germans. It was hauled up by VII Corps troops, who captured Germany's third largest city last week.

More than 2,500 soldiers filled the concrete grandstand and overflowed onto the turf terraces of the amphitheater. Overhead as protection against a German sneak raid were eight P47s of the Ninth TAC which circled on patrol during the 11-minute ceremony.

Maj. Gen. J. L. Collins, commander of this crack Corps, presided, standing in front of commanding generals, assistant division commanders, chiefs of staff and artillery officers of the Corps' four divisions—Third Armd., 104th, First and the Eighth Divs.

Massed before them were platoons of each division and others representing Artillery, Cavalry, Signal, Engineer and Medical troops of VII Corps acting as color guards. The dull grey of this German mid-afternoon was splashed with color from the standards not only of these four divisions but every regiment or other unit who had brought theirs with them. On the standard of the Fourth Cavalry glistened 35 silver rings representing citations from a half dozen wars.

The 104th Inf. Div. Band, under direction of WOJG Fred R. Frey, of Roselle Park, N.J., furnished music for the occasion.

Loralei who used to finish the Lord's Prayer with "the kingdom, the power, and the gory..."

MARCH 1945

AMTLICH!

Der deutsche Soldat, der dieses Dokument findet, hat es sofort seinem Einheitsführer (Kompaniechef, Bataillonskmdr., Kommandeur der Kampfgruppe usw.) zu übergeben. Dieses Dokument gestattet der ganzen Einheit eine ehrenhafte Übergabe. Um nutzloses Blutvergiessen zu vermeiden, haben sowohl Deutschland wie die Alliierten beschlossen, die Bestimmungen des Haager Abkommens von 1907 und der Genfer Konvention von 1929 zu befolgen. Die folgenden Grundsätze sollen daher bei der Übergabe von Kompanien oder anderen Kampfeinheiten als Grundsätze gelten:

1. Truppenkommandeure sind dazu berechtigt, einen Parlamentär mit weisser Fahne zur Besprechung der Übergabe zu entsenden. Dieser Parlamentär gilt als immun.

2. Parlamentäre müssen mit Beglaubigungsschreiben ausgestattet sein.

3. Truppenteile von unter 500 Mann ergeben sich in der üblichen Weise, sollen aber in kleinere Gruppen aufgeteilt werden und sich in diesen kleineren Gruppen den alliierten Linien nähern. Die Gruppen bewegen sich ohne Waffen, Stahlhelm oder Koppel, mit hochgehobenen Händen.

4. Vom Augenblick der Übergabe an gelten deutsche Soldaten* als Kriegsgefangene und geniessen den Schutz der Genfer Konvention. Ihre Soldatenehre wird voll respektiert.

5. Kriegsgefangene erhalten dieselbe Verpflegung wie Angehörige der alliierten Heere und werden, falls krank oder verwundet, in denselben Lazaretten behandelt wie alliierte Truppen.

6. Laut Genfer Konvention dürfen sie weder der Gegenstand von Repressalien, noch der öffentlichen Neugierde preisgegeben werden. Nach Kriegsende werden sie sobald wie möglich nach Hause entlassen.

* Als Soldaten werden auf Grund des Haager Abkommens angesehen: alle bewaffneten Personen, die Uniform oder ein weithin erkennbares Abzeichen tragen.

Germany—March 2, 1945

Dear Mother, Dad & kids,
   Not much time to write but I thought I'd drop a line to let you know I am okay.
   The other day I was slightly wounded in action but nothing to worry about. I was back on duty as soon as they put on a bandage. I got the Purple Heart and am sending it home. All it was were a couple of scratches from shrapnel and I'm good as new so don't worry. The medal really looks nice.
   Have you received my package yet? I'll send another soon.
   One thing I would like to have is some cocoa—I don't drink coffee and I like a hot drink on these cold wet mornings. So please send some if you can along with some other food. Anything to spread on bread goes good. I found a can of caviar the other day and it was swell also some sardines would go good.
   The paper is coming through fine, usually takes about three weeks to get here and I read it from front to back.
   We also found 20 chickens today and when we stop long enough to cook them we'll have a big feast. I shot a Kraut officer the other day and got a brand new pair of 10 power binoculars which are much better than the ones that were issued to me. I'll bring them home along with my pistol after the war. Write soon.

                                                              Love, Jud

July 6, 2014

Dear Dad,
   Just some shrapnel you say. You never told us anything more than that you got the Purple Heart. I found a newspaper article about you written after the war that talks about your comment on receiving the medal, "standing up when I should have been sitting down."

You wrote on the newspaper that they misquoted you but that really sounds like something you would have said.

Travis had a school assignment to write a short essay on "My American Hero" and of course he chose you as his hero. I am so grateful my children had the opportunity to know and love you as much as I did. Travis' paper was so important to you that it was framed and hung on the wall in your den next to your many awards and medals.

The closest thing I ever had to being wounded in battle has to be my injury plagued, unsuccessful career as a jockey. Riding racehorses can feel like a battle, especially when you are breaking babies in the fall.

Without a doubt my worst fall from a horse was at Blue Ribbon Downs in Sallisaw, Oklahoma. It was in November, the track was frozen as hard as concrete and the temperatures were in the teens. I was galloping a green filly without even a registered name. She did some unspectacular attempts at bucking as we started out, and then leveled out. We were hitting a pretty good clip as we came into the first turn. She then did some kind of an inside out trick and shucked me like a seed in a watermelon-spitting contest. I hit the ground like I had been jack hammered. I could not see; I was blind. I groped the ground to try and find the infield to get out of the way of the track traffic. I was sitting there in a panic, hoping for my sight to return. It was a really horrible feeling. I heard someone ride up and ask me, "Kathy, are you alright?" I did not want this person to know I was blind. I said, "Yes, I am just fine." He then says, "Well, why don't you take your helmet cover off of your face?" I was the most joyful person in the world as I ripped the helmet cover off, springing to my feet saying, "Thank God, thank God, thank God I'm not blind!!" He started laughing so hard he almost fell off his horse. This was not the end of this wonderful lesson in jockey gymnastics. He caught this demon bitch filly from hell and she pile drove me into the concrete track twice more. She had my number. I galloped a few more head then headed to the local burrito place. I knew I was now the object of great amusement for the track. As I walked in there was some good-natured ribbing about thanking God and helmet covers but to tell the truth I was so glad I could see it really did not matter about my embattled pride. I ate with the gusto of one just reprieved from the world of eternal darkness. Then I attempted to get out of the booth I was in. There was something very wrong with my lower back and butt. I was not about to give my fellow horsemen more comedy fodder so I waited until all had left, then I eased my excruciating butt and back out of the booth, unable to stand up straight. I was hunched over like a 100 year-old granny. As I shuffled to pay my bill I remarked, "There must be something wrong with your booth.

I was fine when I came in here." I had to crawl on the passenger side of my truck then roll over to get in the driver's side. I headed to the emergency room where x-rays revealed a broken tailbone. Wonderful. First the comedy of the helmet cover draped over my face and now the ignominy of a broken butt. I did not even have the bragging rights of a broken arm or leg where at least everyone will sign your cast in sympathy. Oh no, I had a broken butt. You can't have people sign that. The only place you can be comfortable is sitting on the toilet. I even tried sleeping on it the first couple of nights.

Love, Your girl who did not receive a medal for her broken butt

Germany—March 9, 1945

Dear Mother, Dad & kids,
As you've probably guessed I was in on the drive to the Rhine—it's really some creek.

I've had several letters and a package from you—the package was "number 1" everything was in good shape and very much appreciated—thanks a lot.

I want you to go ahead and get a ring for Patti—I know things are high but the war with the Japs will probably last a couple of more years and this one isn't over yet. I'll be sending home about $100 in a day or so and about $250 the end of the month. So please go ahead and get her a nice ring for me.

The war is going pretty good but we sure had some hard fighting—in one attack alone my platoon knocked out an antitank gun and captured 85 prisoners—pretty good for only 20 men with 5 tanks.

I sent you another package with 2 flags, a helmet, a Kraut camera (something is wrong with the lens but can be fixed), and my Purple Heart. I now have a very expensive Kraut camera I found—it uses 120 mm film so if you can get some please send it.

I've got a new tank but can't describe it because it's a secret—it's swell—but the Krauts are still way ahead on equipment.

Please write often and get the ring for Patti.

Love to all, Jud

P.S. weather same as usual: WET

July 8, 2014

Dear Dad,

What in the world were you thinking having your mother give your girl a ring? I guess Patti nearly became my mother. Did you propose via letter?

I never told you this but when we were in Vegas at Liberty's first wedding I was extremely stressed out because you and Mother were both going to be there at the same time. The two of you had not seen or spoken to each other in over 20 years. You had been married to your wife, June, for about 20 years by now, but only you attended the wedding. I remember we all went out to eat at the Hard Rock Café prior to the wedding. We were all drinking. Mother was attempting to carry on a conversation with you while you appeared to be most uncomfortable. As can sometimes happen when one has multiple spouses and indulges in adult beverages, the tongue can make a slip. You inadvertently addressed Mother as June but she must have been enjoying a buzz and did not take offense. You looked over at me and said, "Well, it is hard to keep the wives straight sometimes." After Mother and I returned to the hotel, both of us pleasantly intoxicated, she asked me what room you were in. I was thinking, "Holy Shit, what do I say?" She seemed bent upon a clandestine rendezvous. I said, "Mother, Dad is now married to June and has to behave like a married man." To which she replied, "Well, I don't care, I had him first." Can you just imagine my angst? Here was my mother who was in her 80's at this time demanding to know what room you were in. You were in your 80's as well and this situation might have been comedic if I was not involved. I held my ground however, and did not subject you to Mother beating upon your door stating, "I had you first." Luckily the rest of the wedding went off without too many more awkward moments.

I wonder why you received a new tank. There is a mystery about you and your tank catching fire. You nearly burned up in it but that is all that I know. In later years you had to be sedated for those scanning machines as a result of the tank fire. You do not say anything about this at this time. Maybe it happened later. You also told Doug and me that your driver ran away. I wonder if the burning tank was the reason why? You never told us the reason.

                Love, Your girl with rights to the "first" mom

Germany—March 11, 1945

Dear Mother, Dad & kids,

Just a few lines to let you know I am okay and I hope you all are too. I got back some of my photos today and am enclosing them. I sent a few prints to Patti but all the negatives to you. Would you please get some prints made and see that she gets one of each. I wrote a few words on the back of each photo to describe it.

Something must be wrong with the camera you sent as most of the photos didn't turn out too good. However I now have an excellent camera and as long as my film holds out I'll keep taking pictures—I've even got a device where I can take a picture of myself.

I sure was glad to get your two letters, Fred, I'm sorry I don't have time to answer you and David with individual letters but they keep us pretty busy both day and night—I sure do enjoy hearing from you both. You were right about my tank, but I have a new different one now and it's really a honey.

That's all the time for right now. Please write often. The weather is still wet but we usually kick some civilians out and move into their homes.

Love, Jud

July 11, 2014

Dear Dad,

I found one photo that should coincide with the guys in your platoon. There is nothing written on the back but you described having 20 men and five tanks so maybe this is them.

It is in the snow and they are gathered around a tank.

My guess is that this was taken after you guys crossed the Rhine but I really do not have any proof. Another mystery to dwell on.

Love, Your girl left with another mystery

Twenty men & five tanks

July 12, 2014

Dear Dad,

I woke up this morning at 4:30 am to Gracie violently vomiting in the bathroom. Yes, I am telling another cat story. Everything she had eaten the night before steamed on the floor. She then meowed and collapsed on the floor next to the vomit. The day before I had stocked up on new kitty litter, 48 cans of Fancy Feast and milk in preparation for my upcoming trip to China. Brenda, who takes diligent care of my house and animals when I am gone, had been instructed as to her special care. I had given her the monthly flea and worm treatment earlier in the week since the vet said it would not affect her condition by making it worse. I had briefly debated keeping Gracie at the vet's while I was gone but knowing how stressed she gets when she goes to the vet decided that this would only increase her misery and hasten her death.

I realized it was highly unlikely that she would survive during the two weeks I will be gone. I didn't want her to die alone so it was time to take away her pain and suffering.

I buried her out front under the Bradford pear tree next to Harley who had been hit by some asshole who broke her back and left her for dead. I now live on an animal burial ground. Song and Yellow Roll, my two horses; my beloved dog, Harley; and now my cat.

I need to go now, Dad. Time to go cry in the shower, to sob myself to sleep, to put Preparation H on my tear swollen baggy eyes. Time to wake up looking for a cat that is no longer here.

Love, Your girl who no longer belongs to a cat

Germany—March 17, 1945

Dear Mother, Dad & kids,

Enclosed is a copy of the Stars & Stripes with a few lines about our outfit. I think that paper contains more propaganda than Hitler's own paper. Also enclosed is the order giving me the Purple Heart.

We had a couple of nice warm days but now it's raining again, I guess they just don't have spells of good weather here.

All this talk about the Air Corps knocking out factories is a bunch of baloney. Coming across the Cologne Plain I saw lots of war plants (tank factories, steel mills etc.) and not one, not one was damaged in the least. They were going full blast when we rolled through. The only ones that were hurt were damaged by our artillery. The Ford Plant in Cologne was in good shape. In one town the town had been flattened by the Air Corps but the steel plant outside of town was untouched—I can't figure out who the papers are trying to fool. Please write often, I really enjoy your letters.

Love to all, Jud

August 2, 2014

Dear Dad,

I guess the copy of the Stars and Stripes was not kept since I cannot locate one. I would have been interested to read the propaganda and see how it compares to some of the hooey we are inundated with today.

I haven't written because Travis and I were on the Viking trip to China. China is so very different today compared to when you were in the Korean War and fighting Chinese. I remember you telling me that you told your superiors that you had encountered Chinese soldiers in Korea and they did not believe you. We were kept pretty busy in such an incredible country, one that I have wanted to visit since I first saw the Great Wall in a View Master at age 9. The flying round-trip is pretty horrible and way too long, which you know first hand from your military excursions to East Asia. On one leg we were in a plane for 14 hours. I told Travis I thought I was developing a medical condition in my buttocks, maybe my previous broken tailbone retaliating. If I ever go again I will take that "slow boat to China" that you always hear about.

We started in Shanghai touring a fascinating archaeological museum called the Shanghai Museum.

Next we headed to Wuhan where we toured their museum featuring the Bronzes from Zhouyuan and artifacts from the Tomb of Marquis Yi of Zeng. The most remarkable pieces in the museum were the bells. The music for these bells was engraved on each bell and ensures the preservation of this ancient art form.

After we toured the museum we were treated to a performance using a replica of these same bells. Also fascinating was the stand they used to place their drums in.

We then boarded our ship where we were greeted with Champagne and orchids!

The champagne was delicious. I appreciated it very much, as well as the never empty wine glass at dinner that night. I appreciated all of these adult beverages to the point of intoxication. Mind you I have not been training for this event for many years and you know how lack of training leads to poor performance. Hence I found myself being helped up the stairs to our stateroom by my son who has trained for these kinds of events. My lack of proper training led me stubbing my left little toe in the bathroom. It was not an insignificant injury and I limped painfully the rest of the trip, no small annoyance, as we had a lot of walking to do.

We set sail, traveling through sheer mountain gorges including the Three Gorges Dam. We saw hanging coffins, the Shibaozhai Pagoda, tried to learn Tai Chi, did learn Mahjong, ate some gross things like eel and fried chicken feet. Travis tried nearly everything. I did not.

We did NOT try some weird old eggs that are considered a delicacy.

Travis was very adventurous and our tour group began to watch his reaction to these new foods before trying them out. It reminded me of that commercial for Life cereal where the kids look and see if "Mikey" liked it before they would taste it. We had a really super tour guide who went by the name of Michael. He had a great code for restrooms, "The Happy Place." He would warn us of "Happy Places" that were not so very happy as well as giving the women great joy when he would announce, "this Happy Place has Western toilets!" This preoccupation for the women with what kind of toilet we would encounter bonded us in a way that nothing else could compare to. This is because Asian toilets are for squatting.

You will be pleased to learn that I finally mastered the art of eating with chopsticks! As one of the women in our group was becoming frustrated with her efforts to master this art of eating she remarked, "No wonder the Chinese are so small, they can't get hold of anything to eat!" Perhaps we should make a law in the West, "Chopsticks are mandatory" until we all lose weight!

Xian was the city I like the best. It is located a few miles from where the Terra Cotta Warriors are. This is one of the most incredible places on Earth as far as I am concerned and they still have much to excavate, including the Tomb of the Emperor Qin, the first Emperor of China. He also started the work on the Great Wall of China according to our tour guide.

After two days in Xian we went to Beijing. We climbed the Great Wall!

Our tour in Beijing included The Sacred Way, a place where the Ming Dynasty Emperors are buried. There is a marble statue of a General on the Sacred Way. He has a mad face that reminds me of that picture of you on a coffee mug that we call "MAD DAD."

We ate Peking duck, which was really good, toured Tiananmen Square, The Forbidden City, and concluded with a performance of the Peking Opera. I actually liked the Opera but I could tell that Travis had his doubts. He has his own unique version he likes to serenade over the phone to me. By the end of our trip he had gone native.

Love, Your weird girl who likes Peking Opera

Germany—March 27, 1945

Dear Mother, Dad & kids,

Sorry I haven't written but I got a leave and just recently got back. I had a 3-day pass to Brussells and had a wonderful time. What a town! I never saw anything like it. The weather was wonderful and in the afternoons I would sit at the sidewalk cafes and drink beer. They have a real fancy hotel reserved for combat officers on leave and was it swell—good food, baths, laundry, and electric lights. They sure have a lot of beautiful women there. At night they have nightclubs that are as good as any I saw in Washington or Baltimore—also as expensive. The bands have all the latest American dance songs and all the people that worked in the cafes spoke English. I got a big kick out of riding on their trolleys (trams they call them), they are the craziest things you ever saw—they just pack them till people are hanging on the outside. The only bad thing that happened was I had my pocket picked on a tram during my last night. They got my wallet but thank goodness I had about half my money hidden away in different pockets—it sure made me mad though because I think it was a Limey soldier that got it—next time I'll keep my hand on my wallet on those trams.

We've got the Krauts running again. I hope it keeps up. I was in Cologne and it's really flat. I've also been swimming in the Rhine—boy was it cold, but I had to get clean before I went on pass. All I've got to do now is wait 6 more months and I'll be in line for another pass.

My French is really coming right along and I'm not doing bad in German either.

Please write often and send some more packages—everything has gotten here in good shape. I sure would like to have that billfold as I don't have any now.

Love to all, Jud

August 3, 2014

Dear Dad,

I am really glad that you got this time away from the front. Imagine swimming in the Rhine to get ready to party in Brussels!! Now that would make a good opening line to land you a hot date! "Hey there, pretty lady, I swam in the Rhine just for you!" Or maybe, "Only real men bathe in the Rhine." Hmm. Maybe a commercial, "Does your man swim in the Rhine? Send him to Cologne for a dip that will refresh, invigorate, and turn your man into a raging Romeo." Ha, ha. Or "Fresh from the front! Immersed in the virile waters of Germany's Raging Rhine!" "I Rhine for you!"

I get it. You cannot really write home and tell your parents and brothers that your new mission was to properly prepare for a venture into debauchery and bacchanalia in the pristine city of Brussels. One can only imagine. Well done, Dad!!

I seem to be getting silly now. Sorry that your wallet was stolen. I guess this pickpocket was unaware of your tank squishing abilities. Perhaps you had an eye on a lady and this distracted you.

Love, Your girl who admires men who swim in the Rhine

Travis' favorite museum artifact

Champagne & orchids

Terra cotta warriors

Terra cotta warriors pit one

"Ussie" on The Great Wall of China

Mad General

"MAD DAD"

Gone native

APRIL 1945

Germany—April 7, 1945

Dear Mother, Dad & kids,

I never saw so much rain in all my life, in the past week not 6 hours have passed that it didn't rain. It's really miserable living outside.

These Krauts are really fighting hard now—some of them are as bad as Japs, they just won't give up. Most of them have a lot more guts than some Americans. The other day a Kraut bazooka hit our tank, by some miracle no one was hurt and the tank wasn't knocked out—but it just gave everyone a big jolt, so my driver gets scared and jumps out and starts running, leaving us with no one to drive the tank. It made me so mad I darn near shot him. He doesn't drive a tank any more. I got the man who shot the bazooka with my tommy gun. I saw him just before he fired but I was putting a new clip in my gun—my ears are still ringing from the explosion.

I sure wish the U.S. would make a tank as good as the Germans—we don't have one half as good. I think if we had better tanks the war would be over by now. It makes me sick to read where Patton says our tanks are better—you can bet your last cent he never had to fight a Tiger or a Panther with one of ours. Please write often.

Love to all, Jud

August 9, 2014

Dear Dad,

I am alarmed to read you were in danger. Was this the time your tank caught on fire? You never really elaborated on the tank fire. I know the incident deeply affected you because you told Aunt Betty and me you had to be sedated whenever you had an MRI. A few years ago while recovering from a serious heart procedure, your mind traveled in and out of consciousness. Do you remember re-enacting

World War II in your subconscious state? One time, the last time, you were in the hospital I came in and asked you how you were doing and you said, "Fine, I am chasing Nazi women." You then laughed and went back to sleep. You had a very pretty, blonde Ukrainian sitter whose accent may have helped trigger this fond memory as you faded in and out of LaLa land. She told me that you would often be talking in German. I hesitate to think what she might have heard you say! Once I heard you say, "Nicht es ferbotten."

You also wrote about the tank driver running away and I certainly understand your anger. That left you a sitting duck. Was he punished for his lack of action?

I really miss my cat. I hope you were not residing in her. I think I will get another cat. Somewhere there is a cat that needs a good home and perhaps a sentient being. For a while there was a chicken on the property that would not stop following my horse Willie and my donkey Waylon. It was weird. I do not know what happened to the chicken, maybe it crossed the road.

Love, Your girl who needs a cat

Germany—April 14, 1945

Dear Mother, Dad & kids,

Well, I think the war is going to be over pretty soon. I think in about a month or maybe two all organized resistance will be finished. There is still plenty of hard fighting but they just can't stop us.

Thanks a lot for getting the ring, Patti said it was perfect. About the money—I ordered $200 worth of money orders (I didn't sleep either) the other day but it just came back as they have some new red tape you have to go through and we've been moving too fast to get it done. I got a pistol the other day and sold it for $25 so as soon as I get time I'll get $225 worth of money orders and send them home.

I guess I can tell you what kind of tank I've got since I've seen it advertised by Cadillac in Life—it's an M24 light tank and much better than my old one but still not half as good as a Kraut tank. Yes, I was in that drive near Paderborn. Your mail is not censored. About cigarettes—the last ration I got was 5 weeks ago of 2 packs. It makes me mad as hell because the Quartermaster gets a pack per day. I've been taking Kraut cigarettes off of dead Krauts—boy are they lousy. You also asked about food—we very seldom get hot chow, once in a while we get one meal a day. The rest of the time we eat K rations or eggs we find. You also asked me to write to Bishop's folks. If I wrote to relatives of everyone that got killed I'd be writing all day & night, besides they wouldn't like to hear what happened—a mortar blew off the top of his head.

I got a swell package from LC Miller—is that Aunt Lil—please send me full name & address and I promise I'll write.

Love to all, Jud

August 10, 2014

Dear Dad,

You probably didn't realize how fortunate your unit was to be given M24 tanks so close to the end of the war. I read that several units did not receive a new tank until after the war when it was too late. The M24, made by Cadillac, was well-liked for its off road capabilities and the 75mm main gun, but you are right that the German tanks still over powered the American tanks. Hey, at least you had a version of a Cadillac! See you were a Cadillac wrecker, too! You used to take us to climb in and out of all kinds of tanks when we were little, something we always enjoyed!

I guess the Germans were better engineers than cigarette manufacturers since you admired their tanks and found their cigarettes "lousy." Your meals were not very attractive and I deduce that you were employing your hunter-gatherer skills to "find" eggs.

As I think about your food I am reminded of a time in Tampa. One Saturday I stayed home while the rest of the family went out. I wanted to test my hunting skills and catch some crabs for a delicious meal. We had not been living in Florida long and could walk to a sandy beach area on a canal where there were all kinds of creatures we had never seen before, including crabs. Off I went on this quest for a meal from nature, devoid of any grocery store contamination. I got to the beach area we had named "The Point" due to its geographic characteristics. Scuttling across the sand were thousands of Fiddler crabs who threatened me with their claws raised, prepared for battle. They were not what I was hunting. I wanted the big blue ones that are good to eat. Here and there they swiftly scampered as I chased, grabbing with a net and tossing into my bucket. I got six of them. I was so proud of my newly acquired crabbing skills!

I returned to the house and began to prepare the pot to cook them in. Mother was a Maine native and knew how to cook crustaceans. She had instructed me on what to do. I was to bring the water to a boil, then drop the live crabs into the boiling water. When they turned red they were done and I was to remove them.

I brought the water to a boil, selected the biggest crab and dropped him or maybe her, since I don't know how to tell the sex of crustaceans, into the pot. It was horrible, horrible. The crab began to emit a high-pitched scream while frantically trying to scramble out of the boiling water. I was no longer the triumphant hunter-gatherer instead I was now the worst, most despicable person alive, a cold-blooded murderer of crustaceans. I began to cry hysterically. What had I been thinking?

Shamefully I buried the crab in the backyard and returned the rest of them to their home at The Point. I never again tried to catch any crabs.

Love, Your girl who murdered a crab

Germany—April 26, 1945

Dear Mother, Dad & kids,

The war has been moving so fast I just haven't had time to write. I guess I can tell you now that our outfit helped clean up the big pocket in the Harz Mountains southwest of Magdeburg. It was tough going most of the time. In our last action there my platoon knocked out two Kraut tanks, one of them a Panther and killed or captured a bunch of Krauts.

Yesterday I went deer hunting with a machine gun on a game preserve and tonight we had deer steak, red wine, white wine, followed by some rare champagne. We also have a Kraut generator we can plug into a house and have lights and radio. We always move into the best house in the area and there are some really swell ones.

We captured an airfield the other day with a bunch of gliders so we got a jeep to tow them and flew all over the place—I went up twice before I cracked up.

I found out today I've been put in for a medal—I hope it comes through but it will probably be a long time *if* and *when* I get it.

Enclosed are some snapshots taken near the Rhine.

I've got about $300 in my pocket but they have a rule now that if you send home more than you earn in a month you have to fill out a lot of papers to prove that you got it honestly, I guess to stop the black market & looting—everyone loots anyway. I'm going to loan $100 to one of the officers who is going on pass to England to get married—you probably say I'm foolish but he is a good friend and passes don't come very often—besides I know I'll get it back the end of next month. As soon as I get the red tape cleared away I'll send you the rest of it.

I think the war is about over and we'll be leaving for the Pacific any day. So if you don't hear from me in a long time I may be on a boat. I think there is a good chance we will go via the States, I hope. The civilians in Central Germany are mostly unfriendly—many of them are bombed out refugees from Berlin, Dresden, Bremen or Hamburg. Those people really hate us. I shot an 11 year old Hitler

Youth the other day who shot a Panzerfaust (bazooka) at my tank—we are running into a lot of Werewolves or guerillas now. Write soon.

Love, Jud

August 12, 2014

Dear Dad,

I found a photo of a tank you destroyed in this action but it is included in a story about St. Lo, which happened much earlier in the war. My Dad is better than your Dad!!!!

I am trying hard to picture your deer hunt with a machine gun. Did you have to dig out a lot of bullets before you could eat your deer steak? War seems like feast or a famine, depending upon your success or lack thereof.

You guys must have thought you were invincible. Here you were just having survived an intense battle and what did you do in your spare time? You went joy flying on gliders! Had you even received any training before you guys did this? Or was this an alcohol-assisted self-taught training course? I wish I knew.

How terrible it must have been to have to shoot an 11 year old. How horrible that he had to shoot you. Sometimes there are no easy choices in life.

I often wonder what Mother's life would have been like if someone had made the hard choice to force her to get help for her mental illness. Instead it was forced upon her after the marriage was destroyed and the future for the two of you gone.

I had just turned 18 and was attending USF. A few days before Thanksgiving I was called to the Dean's office. Shelley, Doug, and Juddy had been put in foster care. Mother was in jail. You and June were somewhere overseas. Mother had gotten into an altercation with a clerk at the local grocery store when she tried to buy cooking sherry on a Sunday, which was illegal at that time. Mother rarely drank but she did use sherry in some of her cooking. They probably thought she was a wino or something. The police were called because Mother was so irrational. It did not take them long to figure out she needed mental help. She was put into the county hospital mental ward. Not one relative came to check on her or us. I was glad because I knew I had to be the one to hold things together for her. She needed us even more than she needed her sanity.

I went to see her; the place was like something out of a horror movie. Everything was painted in this pale, sickly green color. Mother used to call it, "baby shit green." They unlocked a door that led into a large room where the inmates were roaming, moaning, slobbering, and sobbing. They came up to me and made pitiful attempts to talk, heart-wrenching motions for help, and would touch me tentatively as if I was someone so precious and needed. Some were rocking and crying, some were in fetal positions on the nasty floor, and some were incontinent. Such misery I had never before seen or imagined. This was where my Mother was being kept. I was then shown the locked room adjacent to this common room of vacant minds and wretchedness. They let me in to see my Mother.

I do not know how to describe that moment. She was so happy to see me and I was so miserably happy to hold her. Terrible, terrible, stuck in this moment. Terrible, terrible that the water had been too hot that afternoon in the tub and led to my brother's disfigurement. Terrible, terrible that my mother was now in a place that would surely make you mad if you were not already there.

They kept her there for three months. It was a truly horrible, desperate, and hopeless place.

You and I corresponded. I was given custody of Shelley, Doug, and Juddy. The health officials finally arranged for Mother to go to a VA facility in North Carolina. The day she was to leave I accompanied her to a hearing before a judge. She was brought before him trembling and crying. He pronounced her "MENTALLY INCOMPETENT." They took my Mother from me as she begged me to save her and not let them take her from her children. Once again I was the "Little Mother." Once again the Forever Sadness took hold. I can still see that scene play over and over again, "MENTALLY IN-COMPETENT, MENTALLY INCOMPETENT, MENTALLY INCOMPETENT." How wretched life can be.

Love, Your tormented daughter

Donkey Waylon, horse Willie and weird chicken

Mother in her Army nurse uniform

Fourth Cavalry's Part In St. Lo Break-Thru

(This is the third instalment of Fourth Cavalry Group's record of achievements in the European campaigns. In this issue we relate the Cavalry's part in the VII Corps break-through between Marigny and Ste Gilles.)

The period from July 26 to August 1st, inclusive, marked the critical phase of the First Army's St. Lo break-through. Fourth Cavalry Group's part of the drive was primarily to screen the flanks of spearhead forces, holding open the gap and permit the uninterrupted flow of troops and supplies. After the initial break-through it became necessary to deepen and widen the penetration. This task was assigned to the First and Fourth Infantry Divisions.

July 26th to 29th Fourth Cavalry Squadron screened the movement of the First Infantry south and west of Marigny. Troops A and B cleared out the area around Les Herouard and covered the north-south road to Marigny. On the afternoon of July 27th Troop A, reinforced by the third platoon of Company F, moved west of Marigny toward Le Lorey. They encountered the enemy about one and a half miles west of Marigny and after four and a half hours of fighting repelled the enemy from

This picture is from the album of T/Sgt. Zalman Friedman of 24th Cavalry Squadron. The tank is a German Mark V and was knocked out by F Company of the 24th. Friedman is posing by the tank.

a position to cut the road of advance and split elements of the First Infantry Division. Five Cavalrymen were killed in the action and 15 wounded. Third platoon of Troop A lost its leader, 2nd Lt. Lawrence E. Elman.

July 28th Troop C, helped by a platoon from each of Troops A and E and F Company, captured the town of Le Lorey and cleared out the surrounding area. They took 70 prisoners from the 13th Parachute Regiment, SS Hitler-Jugend-Division, 353 Infantry Division and 420 Panzer Grenadier Regiment. The Squadron held a solid screen from Comprond to Marigny.

Group was released by the First Infantry Division July 29 and attached to the Fourth Infantry for reconnaisance in the Hambye and Percy area. Slight resistance from enemy infantry was wiped out southwest of Hambye. Patrols found a line of determined enemy resistance from Villebaudon south to Percy. July 31 the Fourth Squadron began recon-

Granville, Avranches, St. Hilaire du Harcourt, Mortain and Vire. It would play a vital role in the entire VII Corps advance. Fourth Cavalry Squadron was given the mission August 1st.

Reconnaisance discovered the principle enemy resistance northeast of the town. Troop B, mounted, encircled and approached from the west and southwest. Troop B, dismounted, advanced from the north. Swift drives over-riding all resistance swept them into the town. One hour after the first patrols entered the town Villedieu les Poeles was entirely occupied.

In the following six hours the Squadron beat off four determined enemy counter-attacks. Harassed by heavy artillery fire they held the town until relieved shortly before midnight. The Squadron was reassigned to the First Infantry Division August 2nd. Its mission was to screen the Division's right flank on the drive toward Mortain.

24th Screens 2nd Armored

Cavalry Helps in Break - Thru

(Continued from Page 1) on the tanks and assault guns. They were assigned to reconnoiter in advance of the 2nd Armored with Tessy Sur Vire the objective.

On the morning of July 27th the Squadron moved south through territory previously bombed by the Air Corps and established a CP near St. Martin de Bon Fosse. Troop A and the 3rd platoon of Troop F joined the 3rd battalion of the 66th Armored Regiment to form a task force to proceed southeast toward Tessy Sur Vire. A second task force composed of the balance of the 24th and the 2nd Battalion of the 66th Regiment reconnoitered in force south to Villebaudon.

The enemy gave stubborn resistance along both routes. After repulsing repeated counter-attacks the first task force secured a position on high ground south of Le Mesnil Opac July 28. At noon of the 28th the other task force captured Villebaudon.

The Squadron was reassembled on July 29th and assigned to screen the north flank of the 66th Armored Regiment on the drive east to Tessy Sur Vire. A determined attack by the German 2nd Panzer Division accompanied by a heavy artillery barrage July 31 caused the 66th Armored to swing to the south.

This change of direction left a gap that was filled by Troop E dismounted and one platoon of Troop F mounted. The 24th was now covering an area about ten kilometers wide and astride the main highway running east from Villebaudon to Tessy Sur Vire.

Progress was slow due to mines and heavy artillery fire. August 1st the enemy was contacted northwest of Tessy Sur Vire at 1000. At 1630 Troop A had worked dismounted patrols into the town from the north and at 1800 it was entirely in the hands of friendly troops.

The 24th Squadron was detached from 2nd Armored Division August 2nd and reverted to Group control. It bivouaced at Friboux after a forced march of 65 miles and was assigned to screen the 1st Division's left flank.

Purple Hearts and Clusters in Cavalry

wounded on beach, and E. Rich, Herbert Ernest K. Pokorny, tachulski, Frank W. don C. Gaulke, -To-Pieces, Carl A. Kenneth E. Bahnick, Buckley, Sherril R. Sneed, James McDaniel Manning. L. Ballman, Vincent yd L. Grosshickle, S. Staudenmeyer, ulbright, Kermit L gene C. Huffman, Steepro, John M. y L. Hansen, Lloyd eorge Bodack Jr., ardner, William R. e C. Bustillas and erke.

rd M. Olson, Ray-milton, Joseph J. nes L. Gordon, . Abeling, James ngelo Landi, Lloyd ctor Oliva, Thomas Arthur E. Warner, mble, Adam E. aymond Thomas, iddaugh, Clement d Harvey S. Olson. ter I. Victor, John Francis J. Kirby, on, Robert L. Blaha, ackson, Vernon L. H. Beckala, Jerry r V. Debski, Richard

Also Robert R. Lund, Otto L. O'Brien, Gersain Giterman, Roland C. Richards, Carl W. Greenfield, Clayton A. Gunderson, James A. Grant, Roy E. Meeks, Jack W. Benner and Elmer L. Elmore.

Also Filemon L. Valdez, Mario P. Cerame, Raymond J. Fabian, Norman B. Stover, Ralph J. Richey, Robert T. Kubiak, William Barnes, Gerald F. Sherry, Paul F. Collins, Irvin C. Krueger.

This completes the list of Purple Heart holders in Fourth Cavalry Group and the men who earned clusters.

Irvin T. Gritzhandler, William E. Johnson, Robert J. Hogan, Zelmer R. Jones, Willis D. Dalton, Berend Freese, Charles E. Hughes, Kenneth R. Smith, Nelson A. Harvey, Richard Mc Dermond, Cola George, Charles A. S. Hartsell, Stanley B. Doyle, Norman E. Howard and Randolf E. Jennings.

One Oak Leaf Cluster

Major Eugene J. Field, Major Leroy F. Clark Jr., Captain Harry R. Haverstick, Captain Albert C. Sauerman, 1st Lt Arthur H. Stern, 1st Lt Raymond D. Anderson, 1st Lt Wilbert R. Keihath and 1st Lt Barney E. Toms.

Alberda, Cecil E. Porter Jr., William B. Cook, William M. Noll, Robert F. Gerspach, Joe H. Lewis, John K. Wildes, Ernest Bender, George S. Johnson, John Q. Segovia, Lyn Lyman, George P. Boler, Edward F. Shine, Chester W. Palaszewski and James G. Steward.

Also Claude Rice, Bert R. Snelling, William C. Droke, Richard A. Lassman, Carl V. Cooper, Leon E. Kesterson, Thomas H. Lyons, Robert E. Dotson, Glenn H. Kimbrough, Doran R. Parsons, George H. Prescott, Harry J. Zielinski, Robert D. Lowe, Hal H. Garner, Simon D. Sutton and Robert J. Horner.

Also Matthew J. Jasinski, George W. Davis, Albert W. Beck, Robert J. Mevissen, Richard C. Johnson, Joseph J. Gardner, Raymond F. Hamilton, Sherril Morris, Francis J. Kirby, Charles E. Jackson, Victor J. Jackowiak, Willard B. Carmen, Emanuel A. Raffaele, James E. Zimmerman, Emil A. Bukowski, Robert C. Knight, Vincent J. Iorio, Carl W. Greenfield and Gerald F. Sherry.

Two Oak Leaf Clusters

Capt Barrett C. Dillow, Capt Earle E. Fox, James R. Baxley, Edward L. Schnell, James M. Mulcahy, Peter S. Rudowicz,

MAY 1945

Ich sammle Patronen!

Das Geschäft lohnt sich. Ein paar Patronen fallen überall ab, wo nicht aufgepaßt wird: beim Laden, durch schlechtes Gurten, ungezielte Schüsse, Verballern auf Krähen, Liegenlassen im Quartier. Ob sich wohl schon einer ausgerechnet hat, was da so jeden Tag zusammenkommt?

Mein Sammelergebnis:

1 Wenn in einem Abschnitt täglich bei 100 000 Gewehrschützen nur je 1 Patrone und bei 10 000 MG. je 5 Patronen verlorengehen, wieviel Verlust entsteht dann in einem Monat?

2 Wieviel Lkw. sind zum Transport dieser vergeudeten Patronen erforderlich, wenn jeder Lkw. 50 Kisten mit je 1500 Patronen laden kann?

Die Munition brauch mit Verstand —
sie kommt vielleicht aus Schwesters Hand!

„Schmorstiefel" — knorke!

So was riech' ich gern. Die Sohlen sind schon fein „durch", und das Oberleder ist auch bald hin. Die sind nicht mehr zu reparieren! Das wird sich der Träumer sicher nicht gedacht haben, als er gestern abend nach dem Sauwetter seine Trudelbecher auf den Ofen stellte. Dabei sagen sie immer, Leder sei so knapp. Na, mir soll es recht sein!

Was wurde hier vergeudet?

1 Wenn bei 1 Million Landsern, die auf Bunker zu je 20 Mann verteilt sind, in jedem Bunker nur 1 Mann in einem Winter seine Stiefel verbrennen läßt, wieviel Paar sind das insgesamt?

2 Wieviel Rinderhäute sind für diese Stiefel nötig, wenn aus jeder Haut 5 bis 8 Paar (also 6½ Paar Knobelbecher im Durchschnitt) geschnitten werden können?

Wer seine Stiefel schmoren läßt und pennt,
muß sich nicht wundern, wenn er sie verbrennt!

Germany—May 2, 1945

Dear Mother, Dad & Kids,

Well, it looks like the war is just about over in Europe. I'm pretty sure I've seen my last action against the Germans—anyway I hope so.

I've got a new job now. I'm executive officer of the company or second in command. It's a lot less dangerous than being platoon leader and it's a higher position. I now ride in our radio halftrack which is much more comfortable than a tank. Now if my Silver Star would come through, everything would be fine.

One thing I have to do now is be military governor of any towns we stay in and since we don't move around much anymore it's usually a pretty permanent job and it sure is a pain in the neck. We have very strict laws for the Germans such as curfew from 8 til 7 (must be in houses or get shot), civilians cannot go more than 1 kilometer from their homes, no gatherings of over 5 people, car cannot be driven, guns of all types and cameras must be turned in to me, and a million other rules, also no telephones.

I'm enclosing $50 and will put $50 in each of my next two letters and as soon as I get the rest of my money orders. Also enclosed is a "Unit Surrender Ticket."

I sure hope I go home before I go to the Pacific. I think I will but you can never tell.

These Russian slave laborers are the worst people I ever saw. They are more trouble to control than the German civilians. They get drunk and start tearing up things. I had to shoot one the other day. The live like pigs and smell like them. I even like the Krauts better. Please write soon and send some smokes if you can as we get very few.

Love to all, Jud

August 13, 2014

Dear Dad,

Hello "Military Governor." Ha, ha!! At age 20 you were a "Military Governor!" and the job was a "pain in the neck."

Why was there Russian slave labor? Did the German's capture the Russians and use them for slave labor? The first time I read this letter I was stumped but later I came across some letters you had saved from Peter Abate who served with you. You helped to liberate a camp. As best I can discover this was Dora-Mittelbau at Nordhausen where the Germans were building the V-2 rocket. It fits in with where you are on the map. When I was researching I came across information in Wikipedia and lo and behold there was a picture of the very same map that you always had hanging on the wall!

Pete Abate also sent you some pictures. There is one of him in the middle with a mess of fish. He was really good looking!

I wrote to Peter Abate but the letter came back return to sender. I once asked you if you had ever been to a concentration camp. You said yes but the conversation never went any further. I regret so much that I did not try harder to get you to talk about yourself and the war. You did not mention a camp in your letters home. I wonder if this was because of what you were required to censor. For you to have kept these letters from Pete Abate must mean that they were of great importance. I found them to be touching, especially the way the two of you referred to each other, "Your comrade in arms." I can only imagine the strong bond between the two of you. I hope you are ok with me including these personal letters in the book. They speak volumes about how much respect and admiration he had for you.

Pete Abate also sent some photos of you and your platoon. How I wish I could talk to just one of them. So far my quest in that regard is still unfulfilled. Pete did not identify who are in these photos. Maybe once the book is published someone will recognize a loved one and contact me.

What was it like to be a part of this camp of horrors? I have never been to a concentration camp but I did go to a detention center when I studied abroad in Córdoba, Argentina, the summer after you died. There is a detention center, D-2, right next to the most famous landmark in Córdoba, Iglesia Catedral. Many people I talked to admit the Church was complicit in the exterminations of those who dared oppose the regimes. So the irony of a famous cathedral right next to a place of torture, rape, and murder is chilling. During the military dictatorships of the 1970s more than 33,000 people disappeared, mostly through these detention centers. They recreated the Nazi regime; so much for ridding the world of this ideology in WWII. Now D-2 is a museum and a very grim place with the outside wall consumed with giant fingerprints comprised of the names of the missing.

There is also a mass grave in Córdoba that held over 700 bodies. Everywhere you go there you find artwork, graffiti, books, and songs about "The Disappeared." Infants were stolen, their parents killed. The infants were given to those who were in power and then reared as their own. The agony in Argentina continues today as these children, now of age, are learning that those they thought were their parents were in fact those who murdered or were complicit in the murders of their biological parents. When I was there the judicial system was only beginning to prosecute and bring to justice those responsible.

On a lighter note, we went to a chocolate festival in Villa Belgrano. This was a German town but it was in Argentina. It was so weird. They even had trolls!!

They were speaking Spanish but the music playing on the loudspeakers was German. I was astonished to hear your beer drinking song, *Eins, Zwei, Zuffa*!! How bizarre! I felt like I was in the Twilight Zone and halfway expected to find you in one of the beer gardens—it was so surreal. Then all of a sudden the strains of *Lille Marlene* filled the air. It was too much and I began to sob. I thought about all the soldiers around the world and their suffering, longing for home. What else could one do? Weep for you, weep for the futility of war, and weep for what had happened to this beautiful country.

There was one thing I thought you might find interesting about Argentina. They had a woman General who fought for independence. She is esteemed and entombed in one of the cathedrals in Córdoba.

I am empty of anything more to say. I love you Dad.

Your girl who sobs at chocolate festivals

Germany—May 4, 1945

Dear Mother, Dad & kids,

Just a short note to let you know I'm okay—in fact I don't think we'll have to fight anymore here except maybe the underground.

I'm enclosing another $50 and will send more in my next letter.

I don't know why but we very seldom get smokes now—I guess the Quartermaster and the U.S.O. commander or the German prisoners in the States get them. You ought to see how thin some of our boys are when we freed them. It sure makes me mad to hear how they treat the Krauts in the States. Please write often and send some boxes with food—our chow is very, very poor now and we're always hungry when we finish eating. For a country that has everything, we soldiers are getting damn little of it.

Love to all, Jud

Pete Abate in middle with fish

D-2 detention in Córdoba, Argentina-fingerprints of the missing

Villa Belgrano, Argentina-Alpine Chocolate Festival

Trolls in Villa Belgrano, Argentina

Vikings in Villa Belgrano, Argentina

Woman General Jose Maria Paz Córdoba, Argentina

#1

To Brigadier General Judson F Miller USA Ret.
Sir;
 When I recieved the 4th "Cavalry Spur" and saw and read your article about F Troop 24th Recon. Squadron. From 1944 to 1947. I. Said to myself I got to write to the "General", maybe, just maybe? We crossed paths. During the war; in the envelope there are two photos. wich I had in a small Ruvelope, everything else was. in a suitcase and stolen. my. Corp. Book my. pistols. And all else a few pictures SAVED. were in a small Evelope, the fellow you see with the pearl handle 45 pistol is "Hoppy" his driver also in the picture is Perkins. I was his assistant driver. Peter Abate, I send you his picture because maybe? Just maybe? You had some thing to do with and would recall it? "Hoppy" was a Seargeant. And if you notice in the picture he is a Leutenant! He was my Tank Commander,

#2

We had a rough battle taking the city of Nordhausen, wich was almost the end. Right after that Hoppy was made Leotenant, we had a hell of a celebration pinning the bars on him. I joined my outfit just before the Rhine in a small town called "Giesen" first battle with the new M-24's. They laid down smoke we rushed in and picked up a company of infantry men, but what Hoppy and the rest of the tanks did, is we ran out "backwards" or it seemed the cannons were facing the rear but we were going forward at a good speed. Reason Im telling you this General is later they showed us. Picturers, and it looked like we were going one way instead we were going forward, well we crossed the Rhine and the rest is history. One incident that I must tell you about, I dont even remember the town, well what happened over

#3

We all heard another tank, moving along with us, we were the lead tank. Hoppy, told the other tanks to move down the street with us, But! at a given signal "Hand" every body stop short, sure enough. when we stopped! The German Tank couldn't stop fast enough, Now we knew we were being "Stalked" we listned for his track movements. What happened next is "The Gods Truth" I was the "Cannoneer" or we called "Loader." So Hoppy told me Put in A.P. Shells only "Armord Piercing" He told the rest of the Platoon to stay pat" We turned the next street and Lo and Behold there was the German tank it was a small "Mark 4" "Tintera." "Rinny Tintera" was our shooter. I was his loader. "Tinni" was the best shot in the company. "He fired once" The Germans Cannon was shattered just like when you push a cigar against a fellows face. We could not believe it.

#4

"Tinni" sent another shell into the tracks, and it froze! They climbed out and surrendered. But the German Tank Commander! was just getting ~~Dow~~ out of the Tank when "Tinni's" shell hit the track, he went flying about 3 stories up in the air, he had his tank helmet on when he hit the ground but he had to be knocked out, he looked at his men all lined up along side the tank. He turned around and started walking down the street, waving his arms, Hoppy chased him turned him around, took him to his men he looked at his cannon and he was still waving his arms. "My General" we got 4 Lugers out of that one! I must tell you this one more, before I end this letter which I hope made your day more pleasant." And I know it did make mine by just writing to you!

Bud

#5.

Hoppy had always boasted that he had the best shooter in the company which was "Tintera" or I called him "Tinni". When we got the new M-24's 76mm. Cannons, we got all new shells to fit the breach, well right after the "Rhine" I don't even remember the town. There were a row of buildings, the lead bldg. we suspected someone shot a Panzer Faust but missed. So, Hoppy told me to put a new H.E. "High Explosive" shell in the cannon, but set the fuse on "Delayed Action" in case the German was running from floor to floor. Well Sir. "Tinni" fired and we watched, not even an explosion against the building, the brick and stones never shattered. I gulped afraid of what Hoppy was going to say. Well he did! He said "For Chist sake! You missed the building." Tinni looked up at Hoppy and said "...?"

NOTE*
Sir,
Hoppy says the shell went thru a window never touching the bldg.

#6

Tinni said "I never miss" at that very moment the building exploded believe me Sir! Just like you see today when they take out a building in one shot. Right away Hoppy got on the phone to McMann and said, I told you I had the best shooter in the outfit. What happened Sir is that I set the fuse to high. As I said before Sir, I can go on forever with stories. And again I say I hope I made your day I know I made mine by just writing to you.

Yours Truly and Comrade in Arms.
SGT. PETER ABATE
1020 Bow Sprit Point
Lanoka Harbor
New Jersey
08734

"P/S Write if you Like"

Brigadier General Judson F. Miller. USA. Ret.

Dear Judson;
    Recieved your picture and put it in the center of all my army buddies, you were our leader so. Its right on the top. I thought you would like a picture of "Hoppy" Lofgren, when you made him a Leutenant. Along side of him is. Perkins. our driver. I perticuly liked that pearl handled 45 Cal. He carried.. General, I thought for some reason the word came down that Phillip Good Buffalo, was killed. It probably was. Just Rumer's. Im sure I would have a picture of Him? I will inquire from. my buddy. Joe Hagen.

    Ill write again I hope Im not. intruding on your private life.
    Your Buddy Always and Comrade in Arms
    Peter Abate,
" Hopp.L.s Cowwner and Leader "

Dear General Miller,

I went through some of my pictures, and am sending you some, Joe Kush, and Scardella I knew well but I keep trying to place, Good Buffalo, and I can't remember him.

I have never attended any Reunion I have a wife with Alzhiemer and it is a constant job. We are married 51 years. I hope some day may be I'll make one Reunion. We go to Florida in the winter. I have a place in Lake Okeechobee! Good Fishing! This house is also on bay side of the ocean called Barnegat Bay I am right on the water.

I hope some of these pictures bring back memories, I lost contact with "Hoppy" and would give any thing to know what happened to him or where he is.

  Your Comrade in Arms.
  Peter Abate.

Goats, Kush & Joe

Kush & Hollingsworth

Thorn, Urban & Penn

Dempsey hand grenade fishing-its easy

Brigadier General Judson F. Miller USA (Ret) Page #1
4-2-01

Dear Sir;
   I hope this letter finds you and your family in good health.
   Peter Abate here to write a few lines, I am sending you some pictures I found in the attic, with some wedding pictures. I hope they bring back a memory as they do for me.
   I had a slight heart attack 9-10-00 and had 2 stents put in. If I pass the stress test coming up and pass it.? I will have already passed 6 mo. "The Point of No Return"
   I had the misfortune of having my Corp. record book in a carton and it was all stolen carton and all. When I came home years ago.
   I have a new nieghbor who happens to be Jewish. He lives with his daughter. I told her and him that my otfit lead by yourself. Arrived and freed,

CONTINUED.   PAGE #2

A DISPLACED PERSONS CAMP. I SAW THEM WITH THE STRIPED CLOTHES AND VERY SKINNY AND EMANCIPATED;
 BUT I ASK YOUR FORGIVENESS I KNOW IT WAS A VERY IMPORTANT TIME FOR US, AND YOURSELF.
 I CANNOT FOR THE LOVE OF ME AND MY MIND REMEMBER WHERE IT WAS AND WHAT CAMP IT WAS CALLED, I BEG YOUR INDULGENSE IN THIS MATTER, SO I CAN PUT THIS OLD MANS MIND AT REST.
P.S. HE DOESN'T BELIEVE ME.
 I HAVE NO PICTURES, I KNOW WE DROVE NIGHT AND DAY TO GET THERE;
 DEAR SIR; THIS YEAR IF I PASS ALL TESTS I DEFINITLY WILL GO TO OUR REUNION, I CAN'T WAIT I AM 75 YRS. HOPING YOU ENJOY THE PICTURES, ALSO. HOPE TO HEAR HOW YOU ARE. I HAVE A PICTURE OF "HOPPY" WILL COPY AND SEND.
 YOUR COMRADE IN ARMS. ALWAYS.
        PETER ABATE
     1020 BOWSPRIT POINT.
   LANOKA HARBOR NEW JERSEY 08734
P.S. PLEASE SEND ME PICTURE OF YOURSELF.

Peter Howie
1020 Bowsprit Point
Lanoka Harbor NJ 08734-2705

Brigadier General Judson F. Miller.
 Dear Comrade in Arms.
I am glad you liked the pictures
I sent you especially old Joe Kush.
 I spend winter in Florida.
on Lake Okeechobee, I have big
family there we fish all the time.
I am right on the lake. "Big Bass"
I spend summers here in NJ.
on the water I have a home on
Barneget Bay." Salt water fishing.
 We spent a nice winter in Florida.
We left 5-12-AM, bad weather in S.C.
Next day decided to drive through
and come home got home 9 PM 5-13.
We had a sandwich, my wife Franny
went to bed, at 2 AM, she called
for me to help her, she was throwing up.
Her eyes were rolling I called 911. And
they rushed her to the hospital, she
had gotten an anyerism in her brain
and died 8:50 AM next morning.
 Now I am alone, my sister come to
live with me awhile, what I can't
believe. Judson she waited to get home,
 over.

If it wasn't for my family I think I would have cracked. I am glad my sister is staying with me.

I know I was long in answering your letter, and I'm sorry about that.

All the best from your comrade in arms

GENERAL JUDSON F. MILLER & KATHY WILLIAMS

Brigadier General Judson F. Miller, USA(Ret)
8009 75th Street SW
Tacoma, WA 98498-4817
Home Phone (253) 584-4218
Email judd@seanet.com

June 22, 1999

Mr Peter Abate
1020 Bowsprite Point
Lanoka Harbor, NJ 08734-2705

Dear Pete:,

Thank you very much for your recent letter. It was good to hear from you.

I was really sorry to hear of the death of your wife. I know that it must be hard for you and that it helped to have your sister with you.

There is not much news from here. I don't know if I will attend the 4th Cavalry reunion in early September as I have two other reunions late this summer plus visits from my children. In the meantime I play a lot of tennis and golf and try to keep busy

Again thanks for your letter.

Your comrade in arms..

Sincerely,

Your name goes here

PART III

MAY 1945 AFTER THE WAR

Myself and Lt. Loffgren (he was my platoon sgt. and was given a battlefield commission.

Germany—May 9, 1945

Started on 9th but didn't get a chance to finish til today May 11

Dear Mother, Dad & kids,

Well the war is over and I'm okay. It sure seems funny to know the war is finished—at least till we get to the Pacific. They finally told us about the point system for getting out. A minimum of 85 is required to get out. Even if I wanted to get out of the army I don't have enough points. I have one for each month in the army—30, one for each month overseas—12, one for each major campaign—3 (Northern France, Germany, & Battle of the Bulge), and 5 for my Purple Heart—a total of 50 and if my Silver Star comes through that will be 5 more. I don't think they give near enough credit for combat men. Those USO commanders in the rear echelon and in the States had a snap. I'd rather spend a year in the States in the army than lead another attack—you just can't imagine how scared you get but there isn't a damn thing you can do except keep on going. I'm sure glad I'm executive officer so when we go to the Pacific at least I won't be right at the head of the attack.

The magazines & papers are coming through fine. I haven't had any packages in a long time and I could sure use some cigarettes—we hardly get any at all.

It sure made me sick to see all those pictures of people at home celebrating the end of the war and reading about them complaining about rationing. If they think it's rough they ought to come here and take a look. In one of the towns I control there are 400 Russian and Poles in a camp. I gave them 3 horses for food and they were really happy. I really don't like the Russians though, they are so ignorant they are just like a bunch of animals and I don't think they ever take a bath. I even like the Germans better. I sure hope we never have to fight the Russians but I'm afraid we will.

You asked if we were on that march to Paderborn. We were on the left flank of the 3rd Armored Division and covered 175 miles against pretty heavy resistance to take the important city of Brillon. We had infantry riding on our tanks and my tank led the march all the way. It was during this operation that my tank was hit by a bazooka. I was a wreck by the time we took Brillon. Three days and nights of no sleep and plenty of combat. After we took Brillon we took the small town of Altenhieren, just west of Brillon, and after that went east across the Weser and had some rough going in the Harz Mountains above Nordhausen. I had one of my closet calls there when a Tiger opened up and blew up 2 infantrymen right next to my tank. We just managed to get out in time. Thank God I'll never have to go through any of that again. At least the Japs don't have tanks like the Krauts.

I think within 3 months I'll be home for a furlough before we go to the Pacific. One thing that we think is wonderful is the lifting of the blackout. In combat we couldn't ever smoke at night. Now everything is all lit up and it sure is nice. It has been a year since I last saw any lights at night and that was in Boston just before we sailed. Please write soon and get the roast beef and cold American beer ready.

Love to all, Jud

Enclosed are some photos and another $50 money order

August 18, 2014

Dear Dad,

Hallelujah! The war was finally over but you did not say anything about celebrating. How did the American soldiers celebrate the end of the war? Perhaps if you had been in England instead of Germany there would have been more jubilation. There really was no time to carouse the streets because the American forces had to maintain order and clean up the mess left behind from 12 years of Hitler's destructive reign. Now the job was about governing. Hello "Guvner." Ha, ha!

At first when I read that you gave the Russians and Poles 3 horses for food I was aghast. How depraved I thought. Visions of Black Beauty, My Friend Flicka, Fury, the Lone Ranger's Silver, Trigger, and Buttermilk all flashed before my eyes. Then I remembered some of my more memorable encounters with horses. There are definitely some that should be eaten, just not by me.

A Black Sun Bar filly flipped on me while trying to load her in the starting gates. She broke my leg. Liberty, who was 3, was greatly relieved when she saw me, "Oh your leg's not broke, it's still on." Ha! It was broken. I guess she thought I would be carrying my leg.

Silky Hare was a diminutive 3 year-old bay filly we were racing in Pompano, Florida. One afternoon I was grooming her, cleaning her feet, when out of the blue, an escape artist, aptly named Time To Go Sugar, came charging between the shed rows. Silky Hare, startled by a race unannounced leaped straight into the air and came down on my left foot. My foot was pinned under her hoof as I lay writhing like a bug squirming on a shish kabob. She broke all the bones across the top of my foot; I have limped ever since then. At least Silky Hare was a fast horse and did not intentionally hurt me so I would spare her from being the next meal.

However I would have liked to watch someone or something eat Mr. Azure T, who broke my arm. He was slow and as sorry a horse that ever was. The same goes for Etta's Astire Girl, who tried to flip in the starting gates and broke most of my ribs in multiple places. She was stupid as well as slow. I went to see Doc Dean, an osteopath who was sympathetic to injured racetrack people like me. That is right, an osteopath not an orthopedic guy, that is how we did things at the racetrack. He said I had a "flayed back." I think that is osteopathic speak for smushed ribs. He gave me some green pills for pain and a Velcro brace. I do not remember him telling me to make sure to eat food when taking these green pills. So of course I took the pills and ate later. It was a wonderful Bacon, Egg and Cheese biscuit from McDonalds. All of a sudden it was not so wonderful. Let me tell you, there is not a more awful experience than regurgitating a Bacon, Egg and Cheese biscuit with freshly broken ribs followed by dry heaves for 3 hours. That damn horse should be in a McDonald's hamburger.

Love,
Your girl who learned the hard way with horses and green pills

Germany—May 23, 1945

Dear Mother, Father & kids,

   We no longer have any censorship so I can tell you where I am. I am now in the town of Guntersberge in the Harz Mountains. You'll have to get a big map to find it as it only has about 2,000 people. Its about 20 miles NE of Nordhausen. I sure wish we would go home although it's not too bad here. I have a '39 Chevrolet to drive around in and the officers live in the best house in town, I have a whole apartment with radio electric, icebox, etc. The country is fine and there is plenty of good hunting for deer and wild boars. Also right next to town is a large lake with boats and good fishing. It's now warm enough to go swimming. We took over a beer hall and the cinema and have a show every night. We also confiscated all the schnapps (whiskey) and champagne in the district. We have hot showers and laundry and we pay people to clean up our houses and make the beds, etc. This town was scarcely touched by the war, only two or three houses hit by artillery. This is all temporary, of course, since we are just waiting to be shipped to the Pacific or home and rumor has it that we are going straight to the Pacific.

   I found out that I have 4 campaign stars to wear on my ETO ribbon, they are—France, Ardennes, Rhineland, and Central Europe. I dress up all the time now and wear my ribbons (Purple Heart and ETO) and overseas stripes.

   I hope my Silver Star comes through soon, that will really look nice.

   I run the town and it's really a job. The people call me "Der Kommandant." The enclosed letter is one of many that I get from my "subjects." An officer over here really rates, the people all salute you and say sir—what a life. I'll sure be glad when we leave for home (I hope).

I've been working on my Regular Army commission and the colonel says that I shouldn't have any trouble getting it.

Well that's about all for now, please write soon.

Love, Jud

August 22, 2014

Dear Dad,

If the censorship is over I wonder why you did not say anything about the prison camp at Nordhausen or the tunnels where the Germans had been building V2 rockets. Nordhausen was known for housing the prisoners that were not physically able to work at the Dora prison camp where the V2 rockets were being built. Unfortunately, on April 3rd the Allies bombed the Nordhausen prison camp, mistaking it for a Nazi airplane hangar.

Yet your letters always portrayed a sense of relief even when sharing few grim details of war. For example, you chose to write about confiscating all the schnapps and champagne in the district. It was just like you to save the alcohol. Was this following strict Army protocol? How was this reported to Company Headquarters? Was it even reported to Company Headquarters? Hmm. Did you really pay someone to make your bed? You were always very strict about how a bed should be made so I am trying to picture you giving orders to German maids. I also remember you would have a "Dad Fit" if the beds were not up to inspection readiness. Linda, your last housekeeper, would tell me she could not understand why you were so particular about how a bed was made. Little did she know that a well-made bed represented discipline and order.

Love,
Your girl who seeks documentation of alcohol confiscations

JUNE 1945

Guntersberge, Germany—June 12, 1944

Dear Mother, Dad & Kids,
   Well I just got back from a 3-day pass to Paris, I really had a good time. I saw all the famous places, the Eifel Tower, Arc de Triomphe, etc. It really is a beautiful city but they sure take your money. The trip down and back was terrible—15 hours in a truck to Verviers, Belgium and then a day & a night in a beat up train to Paris. We stayed at the Hotel Crillon on the Place de la Concorde (where they used to have the guillotine). My last day there I went to see the Follies, they were pretty good. We took quite a few pictures and as soon as I get them developed I'll send them home.

June 13

Today they had a big parade and the colonel pinned the Bronze Star Medal on me, I got it instead of the Silver Star for which I was recommended. I'll send it home soon. I now have 3 ribbons, Bronze Star, Purple Heart and my ETO ribbon with 4 battle stars. I'm enclosing the recommendation that was sent in, it tells just what happened.
   We also found out today that we are 99% sure of going home, when I don't know, soon I hope.
   Well, that's about all the news, please write soon.

                                                              Love, Jud

P.S. I've got a complete SS officers uniform that I'll send home soon.

August 30, 2014

Dear Dad,

Bravo for receiving the Bronze award for your heroic achievement in battle. You are now in elite company with Ernest Hemingway, Norman Schwarzkopf, Colin Powell, and John Kerry, to name a few. Why are you looking sideways at the photographer as you wait to receive your Bronze Star? I am impressed there was even a band there! Well done!! The best I can do Dad is complete seven marathons, eight triathlons, and receive medals for my endurance.

Sorry I haven't written in a while but with school starting back up and work it has been hectic. It is finally here, my last semester at GSU. I have three classes, Spanish Literature (taught in Spanish), Senior Seminar, and Global Issues. I won't know what to do with myself when it is over. Every night I come home, do some chores, watch the news, and then it is study and book time until I watch a recorded Jeopardy and then off to bed. I will retire from Flash Foods October 1 and that is just around the corner. The book is coming along yet I find that I dread its conclusion.

I have the house for sale. I want to move closer to the kids, hopefully to the mountains of North Georgia.

What a year you have had. I would like to visit Paris one day. I'll bet you and your comrades really whooped it up on leave!

Love, Your girl who has not been to Paris yet

Ermsleben, Germany—June 21, 1945

Dear Mother, Dad & Kids,

Well I am now a company commander. They had quite a switch of personnel the other day and there was an opening for a company commander so the colonel gave it to me. My old outfit is scheduled to be busted up and everyone shipped out except those with enough points to get a discharge. Many of my old men and one of my officers, Lt. Brooks, will be in my company. We are slated to go back to the States *soon*, and the best of all, after our leaves we will train over again in the States, I sure hope I don't have to go into combat again—I had my fill of that over here. I feel pretty good about my new job because I was selected over a bunch of other officers that all rank me. Company commander is a captain's job and if some stray captain doesn't come in and take over I should get a promotion in a few months—it takes a long time when you're not in combat. I am now responsible for over 2 million dollars worth of equipment and over 100 men and 7 officers. I sure didn't expect this so soon.

Enclosed are the citation and order for my Bronze Star Medal and a couple of photos. By the way I mailed the medal today.

My new address is: Company "C," 759th Tank Battalion, APO 339. I now wear the armored force insignia instead of cavalry and the armored force shoulder patch.

My new job keeps me plenty busy so I'd better stop now. Please write soon.

Love, Jud

September 6, 2014

Dear Dad,

I have a full plate with classes and work. It looks like you were also really busy, a different kind of busy. Much better than war, but I guess in its own way the work was very challenging, especially since you were promoted so quickly. I have had a similar experience when I was promoted to an area supervisor. I went from managing one store with twelve employees to supervising eight stores with 52 employees. Later when I became a District Manager I had 7 supervisors, 59 stores and over 400 employees. People can be a real pain in the neck. Our head of Human Resources used to say that 50% of the idiots in the world are our customers and the other 50% are our employees. I could not agree more. At least in your line of work you could court martial or shoot them or something. We have to go through all kinds of hoops to deal with all the BS. I was so stressed when I became a supervisor that for one year, off and on, I would break out in hives. I discovered that running made them go away so I have been running ever since. When I became District Manager I found that I had to add one more ingredient to keep me sane, wine. So I run in the morning and drink wine at night.

Love, Girl who out-smarted hives

Willie & Turk

Kush

1st Platoon

Sgt DeHaven

Gwarzki

Bronze medal ceremony

DEAR DAD

24th CAV. RECN. SQ.
APO §230, U.S. ARMY

27th April 1945

To: C.O., 24th Cav. Recn. Sq. (mecz)
/8
Subject: Rebommendation that 1st Lt. Judson F. Miller, Cav. 0103272 be awarded the Silver Star for action in the Harz Mts.

1. Where---Harz Mt. Area on Roads from Sulzhayn,RC055382 to Crossroad RC053422;thence to Hohegeiss RC-029445; thence to Benneckenste RC-064453;thence to Sorge RC-049481.   in

2. When---Action commenced 1600 on 13 April 45; at Sulzayen; Crossroad secured at 1200 on 14th April 45;Attacked and seized Hohegeiss at 2235 on 14th April 45; Attack on Benneckenstein started at 0200 15th April 45, Town secured at 0800 15th Apr Outposted Benneckenstein all day 15th April and attacked Sorge at 2000; Sorge secured at 2300. Releived at 0630 on 16th April 45.

3. Type Terrain---Extremely hilly, heavily wooded, roads canalized and curving.

4. Atmospheric Conditions---Clear and cool, nights very dark.

5. Visibility---Very limited, impossible to see more than a few yards off to either side of the roads due to heavy woods and vegetation; visibility along roads limited to 50 to 100 yards due to hills and curves.

6. Location of Enemy---Enemy dug in along all roads, armed with rifles and panzerfaust. All strategic points ie: hills sharp turns and crossroads covered by S.P., A.T., or 20mm AA Guns. Resistance was met in all three towns.

7. Enemy Observation---Excellent from ridges and trees, very well organized.

8. Estimated Strength of Enemy-& Character of Fire---One Regt. of Inf. reinforced by Tanks, S.P.'s, 2-105mm A.T. Guns, and 7 20mm A.A. & A.P. Guns. Fire was delivered by all of the above and by small arms and panzerfaust.

9. No.of casualties inflicted and sustained---Estimated 150 dead, unknown injured, and took an estimated 1000 P.W.'s including many Officers. No casualties were suffered by Lt. Miller's Platoon.

10. Narrative---Lt. Judson F. Miller and his 3rd Plt. of Light Tank Company "F", 24th Cav. Recn. Sq. (mecz) was on the 13th April 1945 attched to the 3rd Battalion of the 16th Infantry to participate in the Battalion mission of seizing and holding Hohegeiss, Benneckenstein and cutting the main enemy escape route at Sorge. On 13th April 45 Lt. Miller's Platoon was attached to "I" Co. of the 16th Inf. to seize and hold the crossroad at RC-053422 and to clear the road from Sulzayen to the crossroad of all enemy resistance and strong points. The advance from Sulzayen commenced at 1600 on 13 April 45. The crossroad was secured and Lt. Miller's Platoon held the Cross= roadwhile the Infantry pushed off toward Hohegeiss. While holding at the crossroad Lt. Miller's Section of Tanks was counterattacked by the enemy with one S.P, one Tank and unk.

10. Narrative contd.--: Infantry. The Tank Section under the aggressive leadership of Lt. Miller knocked out the S.P. and drove off the Enemy Tank and Infantry inflicting unknown casualties. During the period Lt. Miller's Plt. held at this point, heavy Enemy Artillery was fired on the position. When the Tank Plt. was relieved from the holding mission at the crossroad by Recn. Troop "C". Lt. Miller led his Platoon forward to continue the attack with the Infantry into Hohegeiss, via a trail paralleling the main road on the right and north side. The infantry advance was led by a Platton of Medium Tanks (5) along the trail through the woods. When the advance elements reached an open spot in the woods at 046434 the Medium Tanks were fired upon by an Enemy Tank hidden in the woods across the open ground to the right. Three of the Medium Tanks were knocked out and the Infantry was stopped. Lt. Miller quickly brought his platoon of tanks to the front and after a quick estimate of the situation, Lt. Miller laid a smoke screen across the open front with his Smoke Mortar, mounted on the M-24 Tank. He screened all movement from the enemy tanks observation and enabled the Infantry, his own platoon of light tanks and the one remaining medium tank to cross the open ground without receiving any further fire from the enemy guns. The combined forces were then able to continue their advance and to accomplish the scheduled cutting of the main road from Hohegeiss to Benneckenstein at RC-043447. Upon reaching this point the direction of the advance was turned to the left and west to seize the town of Hohegeiss. Lt. Miller led the attack on the town and during the action knocked out an Enemy 1/2 Track, 1 Staff Car, 1 20mm A.A.&A.P. Gun and forced two Enemy Tanks to withdraw from the town. An estimated 50 of the enemy were killed. Two of Lt. Miller's 4 Tanks were immobilized at this point for mechanical troubles and were left behind with their crews. One of these was Lt. Miller's Tank but he took over command of one of the 2 remaining tanks. During the quick reorganization for the further attack on Benneckenstein one Enemy Tank and unknown number of infantry attempted to re-enter Hohegeiss but were driven off by fire from Lt. Miller's Tank. At this point Inf. Co. "I" was relieved from action and put in reserve and Inf. Co. "L" was taken out of reserve and put into the action. Lt. Miller was then attached to Inf. Co. "L" to continue the advance. The Tank Platoon now consisted of 2 Tanks. At 0200 15th April 45 the force moved to seize and hold Benneckenstein. The advance proceeded along the main road into the town as planned with the Light Tanks spearheading the infantry movement. They quickly over-ran the poorly organized resistance. Benneckenstein was completely occupied by 0800 15th April 45. Approx. 150 P.W.'s were taken and an estimated 25 of the enemy were killed. The Town was outposted during the day and at 2100 Inf. Co. "L" with Lt. Miller's Plt. of 4 Light Tanks initiated an attack on Sorge to cut the main enemy escape route. This advance led by Inf. Co. "L" and Lt. Miller's Platoon was followed by Inf. Co. "I" with a platoon of Medium Tanks and a platoon of Light Tanks attached. The escape route to be cut led from Braunlaga which was reported, by G-2, to be an Enemy Corps Concentration Point. Braunlaga was to be attacked by friendly units one day after the seizure of Sorge was accomplished. Braunlaga and the escape route were further reported, by friendly tactical air recn., to contain an estimated 50 enemy tanks. Lt. Miller preceeded the infantry advance into the town and over-ran enemy outposts. An Enemy Tank was heard maneuvering in the town. Lt. Miller without firing a shot set up an ambush for the enemy tank to be used at daylight. As soon as daybreak disclosed the Enemy Tank holding the town, it was quickly knocked out.

11. Recommendation—Lt. Miller during this long and continuous action narrated in the foregoing report has demonstrated qualities of Leadership, Aggressiveness and Fearlessness which retain for him the initiative for the attack in his own control. These qualities have enabled him and the men following his leadership to accomplish these several difficult missions.
At any point along the route of this advance and attack, Lt. Miller would have been justified rightly in advising against the employment of his light tanks. His very aggressive action and desire to carry the attack to the enemy, no matter what the odds, so completely disorganized strong points in the enemy's system of defense that the over-running of these organised defences is made to appear easy.
The counter-attack at the crossroad was beaten off by Lt. Miller's Section of 2 Tanks. He did not hesitate for a moment in the laying of smoke across the open ground in the woods and in leading the advance of his Light Tank Plt. through an area where 3 Medium Tanks had just been knocked out. In the counter-attack at Hohegeiss, he engaged the enemy with his own single tank and succeeded in breaking up the enemy maneuver to gain lost ground. Again he did not hesitate to continue the advance from Hohegeiss, though he had only 2 tanks remaining. In the move against Sorge again there was no hesitation to move against reported overwhelming resistance.
It should be remembered that most of the foregoing action took place during the hours of darkness in very heavily wooded areas, not conductive to good tank maneuver or protection.
I therefore recommend that 1st Lt. Judson F. Miller be awarded the Silver Star and cited for bravery and gallantry in action far and above the call of duty.

         ELLIOTT AVERETT Jr. Capt. Cav.
         Cmdg. Co. "F"

CITATION

AWARD OF BRONZE STAR MEDAL

FIRST LIEUTENANT JUDSON F MILLER O1032728 Cavalry, Company F, 24th Cavalry Reconnaissance Squadron (Mechanized), United States Army, for heroic achievement in action against the enemy on 14 April 1945 in Germany. LIEUTENANT MILLER's platoon of light tanks was attached to a company of infantry, when anti-tank fire knocked out three medium tanks just as the elements reached a clearing of about five hundred yards. Seeing that the attack lost its impetus, he ordered the platoon to follow him. He swung out of the column and quickly moved his unit forward to an advantageous position where he laid a smoke screen across the open front, thus enabling the infantrymen to deploy. As the smoke lifted, he moved his tanks across the clearing, delivering withering 75mm and machine gun fire as they advanced. This action resulted in approximately forty of the enemy killed and two hundred taken prisoner, and two tanks destroyed. His courage, initiative, quick-thinking and brilliant execution in a situation that required bold action, reflect great credit upon himself and the armed forces. Entered military service from Oklahoma.

Editor: Fightin Oklahomans

Lieutenant Miller, 20 years old, is the son of Mr. and Mrs. Herbert F. Miller 2165 East 38th Street, Tulsa, and graduated from Oklahoma Military Academy in 1942. He looks forward to early transfer to the United States and subsequent action in the Pacific theater.

JULY 1945

Pancho-real name Emilo Cadierno

Command Post Bauscheim

July 11, 1945

Dear Mother, Dad & kids,

I'm now in a town named Bauscheim which is about 3 miles from Mainz and 15 miles from Frankfort am Main. I don't like this place nearly as well as the Harz Mountains. The towns here saw quite a bit of artillery fire although they are pretty well fixed up now. As usual I have the best house in town, which I share with Lt. Brooks, my executive officer. It's complete with furniture and everything. The hausfrau who used to live here cleans up every day. I'm sleeping between sheets in a huge luxurious bedroom.

The job of company commander is sure a lot of work, you have a million and one things to worry about and just don't have a minute to yourself. One thing that sure seems funny is to have a bunch of officers saying sir and saluting me.

When we moved from the Harz Mountains the Russians moved in and to me the Russians are worse than the Nazis, they are just a wild ignorant, filthy, uneducated mob. You should have seen how they acted when they took over our town, they made me so mad I wanted to open up on them. They sure have a sorry looking army, no discipline, ancient weapons, and just any kind of trucks or horse drawn wagons they could pick up. Also they are very unfriendly to the Americans. I'm very certain we are going to fight them some day soon. The Mongolians are the worst of the bunch, but they are all terrible. They all think Roosevelt is still President and never heard of Truman. I caught one bunch (all drunk on vodka) breaking into a house and tearing up everything and beating hell out of the whole family, I told their officer and the dope said he could do nothing so I told him unless they stopped I'd bring up my 18 tanks and he would have a bunch of dead Russians (Ruskies we call them). Somehow he stopped them. All the Krauts were terrified when they heard the Ruskies were coming, I had one woman, a countess, offer me a box full of jewelry if I would give her a pass to Koln (Cologne). They still haven't said any more about us going home, I sure hope its soon as I'm really sick of this Europe.

I have a Spaniard who goes right along with me who has been with the outfit for 3 months. He really has a history, 28 years old, well educated, wealthy, and speaks Spanish, German, French, Italian, Russian, and what English he learned from us. He was a regular army officer when Franco had his revolution. He fought with the Republicans till they were licked then he escaped to France and joined the French army. When France fell he was captured and escaped and joined the FFI, (French underground) and smuggled American pilots to Portugal, then he was caught by the Gestapo and sent to Buchenwald concentration camp where the SS worked him over. He was there when the Americans came. He is really quite a fellow, we call him Pancho but his name is really Emil Cadierno. I'd sure like to take him home but he wants to go to Spain and start a revolution. By the way, he also hates the Ruskies and communists. Please note my new APO number and write soon.

Love, Jud

September 16, 2014

Dear Dad,
You were definitely prophetic about further conflict with the Russians. They still today behave like a bunch of Neanderthal adolescents with Putin their chief macho, macho man. How this culture produced the composer Tchaikovsky and then a bully like Putin is truly astonishing. I wish I could have met your friend Pancho—I'll bet he told fascinating stories of his life experiences. I wonder what became of him. Did he live to see Spain as a democracy?
Well, today is my 65th birthday. You would think reaching a big milestone like 65 would give one the right to chose how to celebrate. But I had to take the high road one last time and attend my final Flash Foods function on Jekyll Island. I dreaded the whole affair because of my "Tiller the Hun" boss. Of course he was going

to say some superficial BS. The only thing sincere about him is his insincerity. Thankfully, two of the owners were very complimentary about my birthday and my 22 years of service with the company. There were several people who mean a lot to me who tried very hard to make it a good time for me and overall it was. It would have been really great if a rattler or rabid coon had bitten Tiller. One can dream childish things even at 65. I was very professional and you would have been proud. I kept it short and said that "Maya Angelou once said that *you reveal who you are by the way you treat people* and I hoped that I had not disappointed them." That pretty much sums up my philosophy of life.

Love,

Your girl who still has childish vengeful thoughts even at age 65

AUGUST 1945

Bauscheim, Germany—Aug 7, 1945

Dear Mother, Dad & kids,

I was listening to the radio today to a program with a bunch of Waves on it and guess who I heard—Anna Virginia Miller, I nearly fell over.

I saw Major Billy Phelps the other day; he is a rear echelon commander with 12th Army Group, what a racket. I also saw Lt. Paul Tative and I found out that Johnny Ressler and Dixon were both killed with the 2nd Armored Div.

I have a big motorboat now and I go out on the Rhine almost every day. I think I'm getting a leave to England in a week or so and plan go see Phyllis and her husband and also go to London. I'm certainly not going near that hole called Glastonbury.

We have so much champagne that I drink it instead of water with my meals. I'll bring some home as its so cheap here and we buy it with money we took from dead Krauts. My magazines are getting here okay but the paper has slowed up. I really do appreciate them.

Sundays are real nice because we don't have to work and the weather is so nice. We usually take the boat and a bucket full of iced champagne and go out on the Rhine. Cognac is very plentiful and I took over a café and made it into a club for my company where my men can go and drink champagne, cognac or beer and bring their girls and dance. They are really glad that they can fraternize, although everyone was doing it before.

Pancho (My Spaniard) is talking about going back to Spain and starting a revolution, but he is afraid they would pick him up as soon as he crossed the border and execute him. He has invited me to live with him in Madrid when the war is over—a trip around Europe would be pretty cheap for me with all my European friends. Please write often.

Love, Jud

September 20, 2014

Dear Dad,

The spoils of war. You had a boat and you blew off some steam on the Rhine!!

I remember when we moved to Tampa one of the first things we got was an 18 foot Seabreeze with a 35 horsepower outboard motor, as well as skis. You taught us all to ski. One of the more memorable family outings was participating in the boat flotilla during the pirate invasion at the Gasparilla Festival. Mother was in the hospital having a biopsy done but you were a brave soul and took all four of us. The weather that February was stormy and grey. We launched out of MacDill and I remember the water around that point was really turbulent and scary. Once we found calm waters we were surrounded by drunken men driving boats of all shapes and sizes, most of which were three to five times as big as we were. There were some choice words spoken and even with a drunken slur the meaning was clear. It was great, exhilarating, terrifying fun!!! Mother would have been appalled. You thoroughly enjoyed the mayhem and I must confess I did, too. We had a few close ramming calls and many a wake nearly swept us asunder but by dark we had made it back to the refuge of MacDill. You always knew how to show us a great time!! What fun you were!! Always great fun.

                          Love, Your girl who always had fun with her Dad

Aug 11, 1945

Dear Mother, Dad & kids,

Well the news is certainly good tonight with the offer of Japanese unconditional surrender. We just finished a big party (its now 4:30AM). Capt. Averett came over and we really had a time talking over old times.

I got my 7-day leave to England and leave on Aug.18. I plan to go to Boston first to see Phyllis and then to London and spend one day in Glastonbury just to see what the place looks like now. I haven't had a real vacation in a long time and it sure will be nice. They say England is more expensive than Paris but I plan to take it pretty easy. I was hoping that I might see a little combat against the Japs as it would give me some more battle experience, although I dread the thought of ever going into combat again. I honestly believe though that we will fight the Russians within 10 years and that will really be a rough one, much worse than Germany.

I now have 3 pistols that I am bringing home, a Lugar, a .32 caliber automatic and a US .45 that I carried all through combat. I've had a bunch of them but traded them or sold them.

Yes mother I have my fingernails in good shape.

When I go to England I'm afraid I'll have to go by boat instead of plane. When you go by boat it takes about a month, but I'm glad to be away for awhile. This job of company commander has too many responsibilities that I'm about to go nuts. I'm even getting a divorce for one man. I am now known as the "Old Man" and I've got men who are over 45 years old, what a laugh. I'm kind of wondering if Phyllis will even remember me, don't worry I won't wear out my welcome, but I'm sure looking forward to just sitting around in Boston. I've got a couple of very fine bottles of French champagne that I plan to take along to reduce my expenses in England. In London a bottle of Scotch costs about 2 pounds ($8). Tell David to be sure and not to get in the infantry, anything is better than that. I've gone into combat with infantry when they had 40% killed and the tanks had no casualties whatsoever.

I'm really looking forward to coming home, it sure will be wonderful.

How is Mr. Jones (or Lt. Jones) getting along? Tell Mrs. Jones and the kids hello.

Please write often as I really enjoy your letters. Tell David & Fred to drop me a line.

Love to all, Jud

September 21, 2014

Dear Dad,

It is comical to think of you being addressed as "Old Man" when you were only 20 years old. Did the ravages of war make you look older than your age? Or was it all the drinking and sunning on the Rhine? Why was Mimi worried about your fingernails? What in the world is that about? Come to think of it I never saw you with dirty fingernails.

I remember the time in Tacoma when you were suffering from gout; June gave me $45.00 to take you to the Nail Spa by Hunan Gardens. It was operated by very pretty, petite, Vietnamese girls. At first you were opposed to the idea of a pedicure but after both of us harassed you to exasperation finally you caved and off we went. Your feet were so swollen that we went and bought shoes a couple of sizes larger just so you had something to wear. The Vietnamese girls were all business and soon had your feet in the whirlpool while you reclined deep into the massaging chair. What a sight! The pretty, teeny girls just jabbering away, your eyes closing, as you immersed yourself in some well-deserved nirvana.

Love, Your girl who took her Dad to his first pedicure

Living it up on the Rhine!!

SEPTEMBER 1945—HELSTERBACH, GERMANY

The castle is square with a courtyard in the middle.

The jeep with the top is mine — pretty slick huh.

22 Sept. 1945

Dear Mother, Dad & kids,

Well it looks like I'll be over here for quite a while yet. I'm now company commander of Co. F which if you remember is my old outfit. The 759 is going home but since I want to stay in the army I transferred here. This is a regular army outfit and my best chance for a permanent commission is to stay here. I've put in my application for a regular commission and think I have a pretty good chance. I wish you would try a few connections at home such as congressmen and maybe Capt. Hamilton knows some big shot in Washington. This is very important so please do everything you can and let me know what happens. I am now definitely in the army of occupation. I'll probably get a leave home but it will be some time yet.

Please start sending my Xmas packages as I'll be here for Christmas. Also I wish you would send one or two packages every week as we still get nothing in the way of luxury items, in fact we get less of some things now than we did in combat. Please, please, please send lots of smokes—last week I paid $30 for a carton of black market cigarettes. Also food such as sardines, canned chicken, a bottle of ketchup, etc. Also an officer's forest green shirt size 15 1/2 x 32 or 33 and about 9 pairs of crossbars (X) and some 1st Lt's bars. And I'd like to get some white underwear shorts size 32 and shirts size 36. I'll send the money for all of this stuff and I sure would like to have it as quickly as possible.

I didn't go to England because I couldn't get a plane and now I'm too busy. I can get a leave anywhere in Europe anytime I want it but I can't make up my mind where to go. I think it will be the French Riviera or England.

I now live in a huge beautiful house on a high bluff overlooking the Main River. I'm between Frankfurt and Mainz.

This will not be our permanent occupation zone, we'll probably go to Heidlberg although there is a rumor we'll go to Berlin. Heidleburg would be okay because it's the only big city in Germany that is totally undamaged by the war, it's really a nice place with good beer and lots of cafes. Where I am now is real nice and I have my 3 motorboats right at the house—1 40 ft cabin cruiser and 2 smaller speedboats. I sleep till about 9 every morning and the work is beginning to ease up quite a bit. The only thing I'm worried about is that pretty soon they will be bringing in a lot of Regular Army officers into the squadron, that's why I'm anxious to get my regular commission as soon as possible. The colonel said he would do everything possible to help me so I think with that and if possible a little pull back home I should make it okay.

I'm getting where I can speak German pretty good so when I come home Dad and I can carry on a conversation in "Deutsch."

You asked if I had my dress uniform. I had to buy a lot of new stuff because my old blouse and some other things burned up when my tank was hit by a bazooka back near Paderborn. I talked to a major the other day and got the straight story on how General Rose was killed, it sure was different than what we read in the papers.

I'll try to write more often.

Love to all, Jud

New address
Co.F, 24th Cav.Rec. Sq. APO 158
c/o PM, N.Y.

September 27, 2014

Dear Dad,

I am so sorry you weren't able to get to England and see your friends. You deserved to have that time away from your unit. At least things slowed down in Germany. I can't believe you did not get out of bed till 9 am. What kind of military order and discipline is that? I did not know you could even sleep that late in the morning. You never did that as our Dad. But then again if you were staying out and partying all night it might have been a challenge to get up early.

I got another cat. Yes, Dad, another cat story just for your enjoyment. I think retiring, finishing college, selling the house, and writing this book are all driving me nuts because I started breaking out in hives again. If I run at least 4 miles I usually don't have too bad a reaction but sometimes a stressful event will set me off. What is really distressing is that these hives usually manifest themselves on my face so I look pretty hideous, which adds to my anxiety, etc. So I thought that maybe getting another cat would help. Pets are supposed to relieve stress. So I figured, run 4 miles or more every morning, pet a cat a lot, and drink wine at night, and the dreaded hives will go away.

I went to the shelter and asked to see female cats. I picked out one that was 5 months old, already fixed they said, complete with up to date shots. They said you could tell if the cat had been fixed by looking at medical records and if the ear tip was snipped. This cat named Bubblegum had a snipped ear so I took her home. Then I headed straight to my vet to have him look over the cat once again to make sure no shots were impending. He takes one look at the cat and tells me it is a male but he has been fixed. I didn't look at the cat's private parts! (or lack of them). I felt really stupid; in my defense the cat is really fluffy. He is black and white and I have renamed him Ninja since he can really sneak up and scare the bejesus out of you.

Love, Your girl who belongs to a cat again

OCTOBER 1945

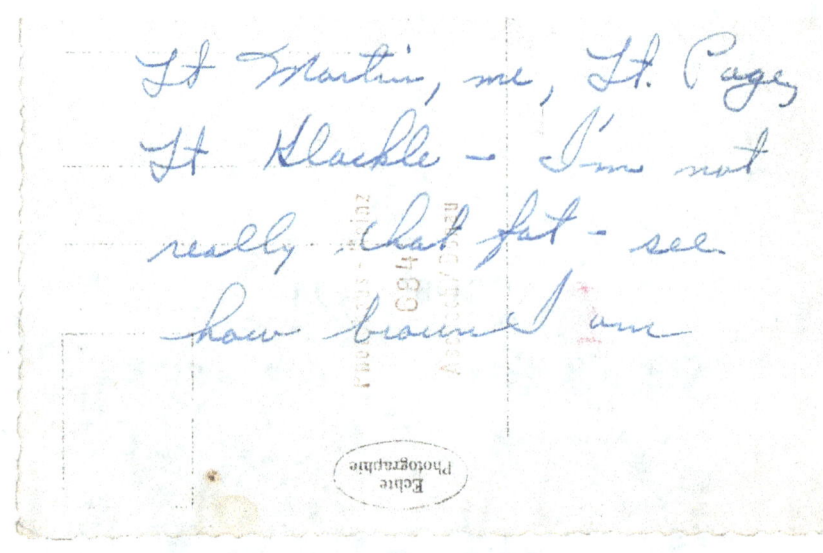

Ginsheim, Germany—2 October 45

Dear Mother, Dad & kids,

The weather is starting to get plenty cold and I don't like it a bit. The one nice thing is that the French Riviera is just like Florida and I can get plenty of passes. I turned down one yesterday as I have so much work organizing my new troop with all the new men.

I bought an excellent camera yesterday for $150, I'll take some photos, then take it to France and sell it for about $400 or $500.

Sunday I was up on the Main River in my boat and passed Gen Eisenhower in his boat. I saluted but he was holding a drink so he just waved and said, "Hi there." You ought to see his boat. He sure is different than most of these crabby old colonels & generals.

I got a letter from Capt. Averett, he is in Le Havre on his way home, which is in Chatham, N.J.

About my letter writing, I never was very good on my correspondence and now with all this reorganization it really is difficult to write.

Patti hasn't written in so long I'm just not going to write till I hear from her and if I don't get a letter pretty soon I'm going to get my ring back. I'll be darned if I'll have her wearing it and running around with a bunch of 4 F's to silly sorority parties and the like.

I saw the recommendation that the colonel gave me on my Regular Army application and it was much better than I expected. He rated me number 4 of all the officers in the regiment. I found out why I haven't been made a captain yet—you have to be a 1st Lt 9 months and a company commander 3 months before you can be promoted. I'm now eligible and I think I'll get it one of these days. The major gave me all his captain's bars and said to, "hold on to them for use in the near future."

I haven't been getting the paper lately and sure miss it. Please see about it and see if I can get the daily as well as the Sunday edition. Also magazines—Esquire, Liberty, Sat. Evening Post, Cosmopolitan. I have nothing at all to read and since the war is over they should be able to send them here.

We are having a little trouble with Germans cutting our telephone wires and stuff like that. In the Harz Mountains we stopped that by burning down a house every time something happened.

I am now in Ginsheim, which is right on the river, and I have a beautiful house. We expect to move within a month and everyone hopes to Bavaria or Austria.

Please write and let me know how everything is at home since the war is over, also how are you are making out about my regular commission.

Love, Jud

PS In addition to those other things I asked for in my other letter please send some 120mm film.

October 1, 2014

Dear Dad,

Today is the first day of my retirement from Flash Foods. No more robberies, no more 24/7 phone calls, texts, emails, and best of all no more Tiller the Hun.

The hives are better but not altogether gone. I still have to finish college. How different this final semester is from my first one when I was 18.

My first college experience had me "the poor starving college student." I would sometimes wait for the Student Union to close to get the leftover stale French fries that the angel at the food concession would give us. Boy, they sure tasted good with a sprinkling of vinegar and a squeeze of ketchup. No microwaves back then. One of my most delicious meals cooked in my termite-infested apartment was the last of my larder, onions fried in real butter. I sometimes stole toilet paper from McDonalds and I hitchhiked to class. It was actually one of the greatest times of my life! I just didn't know it at the time. Then Mother's arrest came and I moved back home. I wrote letters to keep you informed on our family situation, you wrote back with words of encouragement and guidance. The transmission went out in the car and Doug broke his leg. College collapsed but I held things together for Mother. When she came back she was a changed woman, got her driver's license back and went back to nursing, something she had given up to follow you and raise a family. I was so inspired by her struggle for normalcy. Too bad that she later slipped back into her madness, disappearing inside of herself while emerging unpredictably. No matter what though, she would always tell me as I went out the door, "Do your best." I wish my best had been better.

Now I have $150 worth of Panera Bread gift cards, a gift from my "Monday Lunch Bunch." I can afford toilet paper and I have a car to drive the 2¼ hour one way commute. There is no Mother to tell me to "Do my best," and you do not write.

Love, Your girl who stole toilet paper from McDonalds

Ginsheim, Germany—11 October 45

Dear Mother, Dad & kids,
   Well I'm going to England for a 7-day leave starting Oct. 13. I get a train at Frankfurt and go to Paris where I change trains for Le Havre, then a boat to Southhampton. Then a train to wherever I want to go. I plan on looking up Phyllis and spending a little time in London.
   I see where they finally caught up with Gen. Patton, I never did care much for him. The 1st Army was a heck of a lot better than the 3rd.
   If I can I'll call you up from England but don't count too much on it. If I call I think it will be in the afternoon.
   The way these Europeans drive a car scares me to death—all they seem to have on a car is a horn and an accelerator. When they come to a crossroad they start honking their horn and step on the gas. They *all* ride bicycles—right down the middle of the road.
   It's getting too cold to go out on the Rhine anymore. About all there is to do is sit around at night or go to the parties that go on two or three nights a week. I wish we would move to our permanent occupation area so we could really settle down.
   Enclosed is a copy of our weekly newspaper and a list of all the officers and their home addresses. Please write soon.

Love, Jud

October 15, 2014

Dear Dad,
   I got my first rejection letter for our book in the mail today. I guess that makes me a "bonafide wanna be author." I am not sure what that makes you.
   It sounds as if you were actually getting bored with parties and

peace. Is this when you were chasing Nazi women? I guess you would not write home about that.

When I was in Argentina I experienced some crazy driving. Some of the intersections in Córdoba do not have a stoplight, only some mirrors (I am not clear on how they used these). Those cab drivers must have some kind of built in sonar or radar to be able to drive like they do and not get killed. I never even saw one single wreck while I was there.

I guess I will pour a glass of wine and celebrate my rejection letter; maybe it's close to our "happy hour" time. Here's to you, Dad!

Love, Your girl who celebrates rejection with wine

NOVEMBER 1945

Lauterbach, Germany—Nov. 13, 1945

Dear Mother, Dad & kids,

    Sorry I haven't written but I've been pretty busy since I got back from England. I had 10 days there and it was really wonderful although the trip was a nightmare. I spent 7 days in London and 3 in Boston with Phyllis and Basil. I got a train from Frankfurt to Paris, changed there for one to Etretat just below Le Havre, it's a famous resort town and we waited there till we could get a boat from Le Havre. Left Le Havre at 10 am and arrived in Southampton about 5 at the *very same* dock I left from to go to France in '44. Got a train from there to London arriving about 8, after I ate and cleaned up I was ready to go to bed. Stayed at the Jules Officer's Club just off Piccadilly on Jermayn Street. (that was Friday) From then till Tuesday saw all the important places—Parliament etc, and went to all the night clubs. I met a R.A.F capt. on leave and ran around with him. He knew some girls and we really had a time. Monday morning I had my billfold lifted on a subway—I still had about 6 lbs. in my shirt pocket but that's not much. I wired you all that day but never got the money. I guess Dad knows how to go about getting it back. I called Phyllis Monday and went up on the train Tuesday. Basil met me at the station in their American Ford (Phyllis was cooking supper, can't get a maid). They really treated me fine—don't worry I behaved myself. Basil is really a fine fellow, he has a big factory that is really something. Phyllis looks fine—she and Basil sure get along well. They have a beautiful house and two Scotty dogs—the black one is naturally named "Mischief" after the one she had in Tulsa. Tues. night sat around and talked and went to bed. Wed. afternoon Phyllis and I went to the show and then Basil took us all out to dinner at a place resembling Petroleum Inn—didn't look so hot but the food was swell. Thurs. went down and looked at Basil's factory—that night Basil and I went out alone to one of the pubs then came home and talked a while. Friday morning got a train to London. I'll never forget what a lovely time I had with them—it was really nice. When I got back to London I thought my money would be in but it wasn't so I sold my camera. Continued to have a good time till Sunday then took a train to Southampton, saw the sights and caught a boat Monday for Le Havre. It was a slow trip because of all the floating mines that the storm had caused. Stayed a couple of days in Paris and then came back. Found that the 24th had moved to Lauterbach, just about 50 kilometers east of Marburg.

I now have command of "B Troop," they gave it to me because I used to be in a recon. troop in the 42nd Div. I like it about the same as "F"—same address except B instead of F now.

We leave here Nov. 21 for Linz, Austria to take up the mission of patrolling along the Danube River which is the boundary between the Russian and American zones in Austria. That's really a good deal as the Austrians regard the Americans as liberators and they say it's really nice there. The 42nd Div. is in Salzburg, Austria so I'll try and look up Stone and Green.

I got your package with the shirt—thanks a million folks, I really appreciate the things you sent. I'll send home about $175 the end of this month. One thing I would like to have is some silk pajamas and a nice bathrobe, I'm going to live in luxury this winter.

It's already snowing here and looks like a cold winter.

Sat. night we had a German concert here for civilians. I decided I needed some culture so I went, it wasn't as stuffy as I thought it would be. I met one of the fraulines (name underlined on program) and took her out to the club for our Sat. night dance. She speaks English like a lot a Germans and comes from Wiesbaden.

As for Patti she can go straight to hell—I didn't want to break up and sure didn't expect it, but now I'd marry one of my Kraut girlfriends before I'd speak to her again. I guess it's too embarrassing to be running around with a bunch of home front commandos and wearing a ring too.

Thanks a lot for working on those congressmen for my commission, I believe it will help a lot—this army is 90% politics.

I found out something the other day I wish I had known sooner, one of the officers from F company who was badly wounded by a

----------in the leg was and maybe still is at Boston General Hospital, Chickasha, Okla. His name is 1st Lt. Browning.

Well the next letter I write will be from Austria between taking shots at the Ruskies (Russians). Please write and thanks for the package.

Love to all, Jud

PS I lost Dave's address, please send it and I'll write.

November 10, 2014

Dear Dad,
This is the second time you had your wallet lifted. That really stinks. How ironic that you came back to the very same dock that you departed from. On top of that Patti breaks up with you. Not the best time for you.

I was most impressed however, that you went and found some culture. That is weirdly funny to me, as Mother was the one more into Mozart. She could play numerous musical instruments with her forte being the piano. When she was in a nursing home and kind of in and out of things, she would still play the piano. I remember you were so happy to hear that I got her piano after she died. I was so touched by that, it just felt so right.

One of the oddest moments that you and I shared was your last Christmas. You had been very sick and on that Christmas Eve you were having trouble sleeping. On the nights you could not sleep you would go to the den and try to slumber in your chair. I would sleep on the floor at your feet. You were flipping channels and you came upon a performance of The Nutcracker. Nureyev was one of the stars. You and I watched until we both fell asleep. As far as I remember you and I had never gone to a ballet or classical concert. This was our first. It was also our last.

Love, Your girl who shared a moment of culture with her Dad

3. Großer
Opern- und Operettenabend
Beliebte Arien, Duette und Rezitationen

Mitwirkende:
Marie-José Cepernic, Opern-und Operettensängerin
Edmund Lenker, Lyr. Bariton, Stadttheater Regensburg
Katrin Burghardt, Rezitatorin, Wiesbaden
Orchester Vollmöller

1. Ouvertüre aus „Figaros Hochzeit" . . . . Mozart
2. Arie des Cherubin „Ihr, die Ihr Triebe"
   aus „Figaros Hochzeit" . . . . Mozart
3. Aus „Zauberflöte": . . . . Mozart
   a) Papageno-Arie
   b) Duett „Bei Männern, welche Liebe fühlen"
4. Rezitation: „Der Zauberlehrling" . . . . Goethe
5. Allegretto und Menuetto a. d. Sinfonie Nr. 11 Haydn

Pause

6. Orchester „Orpheus", Ouverture . . . . J. J. Offenbach
7. Rezitation „Entstehung des Kusses" . . . Eugen Roth
8. Walzerlied . . . . Schröder
9. Duett aus „Hochzeitsnacht im Paradies" . Schröder
10. Melodien aus „Fledermaus" . . . . . . Strauß
11. Rezitation „Klein Lieschen" . . . . . . Eugen Roth
12. Aus der „Maske in Blau", „Die Julischka" . Raymond
13. Duett aus „Gräfin Mariza" : . . . . . . Kálmán
    „Komm mit nach Varasdin"

Lizenz Nr. 5013
Veröffentlicht unter der Zulassungsnummer 5013 der Nachrichtenkontrolle

Concert date with Fraulein Katrin Burghardt

DECEMBER 1945

Willie at motor pool in Bauscheim

Scharding, Austria—5 Dec. 1945

Dear Mother, Dad & kids,

Sorry I haven't written but since we've been in Austria I've been working pretty hard and was sick in bed with the flu for 4 days, I still don't feel so hot but there is so much to be done I just had to get up.

Today is my birthday and it sure is a lot better than the one I had a year ago in the Hurtgen Forest. I got two more packages a couple of days ago and they sure came in just right for my birthday. Thanks a lot folks, I really appreciate them.

Austria is positively the coldest place I've ever seen. There is about a foot of snow now and the natives say that is nothing to what is coming.

What makes me really sore is that we have to treat these people as allies, can you beat that? The Austrians hate the Americans worse than the Krauts, yet we have to be very careful not to offend them. I think General Mark Clark must have been born in Austria the way he loves these people. To top it all the name of the street I live on is "Herman Goring Strasse."

If you have a map of Austria you can probably find this town, Scharding. It's on the Inn River in the northern part of the American zone in Austria. We are now patrolling the Danube River between Passau and Linz with the Russians on the other side. They sure have a lot of Russians over there compared to the number of troops we have here, I wonder why.

I have about $150 I'm going to send home but I have to drive to Linz to get a money order and its about 140 miles round trip which is quite a drive in this snow. As soon as I get time I'll go down. Enclosed is 5 Austrian Schillings, worth about 50 cents.

The chow we get here is very poor, not much of it and all dehydrated and canned goods. It sure seems funny that we can't get good chow now that the war is over. I guess all our "allies" are getting it. Smokes are also hard to get and the only ones we get are Phillip Morris.

It looks like I *might* come home sooner than I thought. There is only one officer ahead of me and then I go to the States for 30 days. I hope I get my captaincy before I go. I don't think I'll call or anything but just take a taxi right to the house and give you all a surprise. Boy it sure will be good to get home again. I hope I go by plane instead of boat, that will really be nice. The only thing I'm afraid of is that I'll get snowbound here. We drew 30 days rations in case we are snowbound.

In this area are tens of thousands of DP's, Russians, Hungarians, Yugoslavians, and every other nationality who came here to escape the Russian Army. About half the Hungarians claim to be counts or something.

Please send me a copy of the Tulsa phone book, I'm going to try and write a few more letters.

Well Merry Xmas to you all, and you can be sure I'll be thinking of you and wishing that I was there with you.

Thanks again for the packages and please write often.

Love to all, Jud

December 5, 2014

Dear Dad,

Happy Birthday. You would have been 90 today if you could have hung around a little longer. It looks like life after war was a great deal different. It is hard to believe that no one had torn down the "Herman Goring Strasse" sign.

I have finished putting out my Christmas decorations. As I was decorating I recalled taking Mother to see the Rockettes in Atlanta for her Christmas present. I had bought tickets for a Sunday afternoon performance. We first enjoyed lunch at the Hard Rock Café then we parked at the bottom of a hill behind the Fox Theatre. She was huffing and puffing as we made our way to the lobby, complaining all the way about having to hike up the hill. As we waited in line there was some mild bumping and once again she let her ire be know with a growling, "In a hurry girls?" upon each bump. As we climbed to our seats she was muttering something that sounded like obscenities. In truth the seating was very steep. Finally we made it. Her first breathless remark was, "I'm going to need a seatbelt." Fortunately those around us thought this very funny and offered to catch her should she begin to fall towards the stage. I was just glad her remark was free of obscenities. By this time I needed some wine and with Mother chatting amicably with those around us I went for it. All I would have to worry about later was how to navigate the downhill hike without being too wine fuzzy after the show. The show was everything I hoped and she really loved it; that is until the very end. The end of the show has a manger scene with live animals, incredible costumes and scenery yet it was missing one vital thing; there was no baby in the manger. Perhaps they were afraid to offend someone. Well Mother blurts out quite loudly (perhaps the wine increased her volume), "Where's the baby? At my church we always have a real, live baby. They can go to Grady and get a baby!" (Grady Hospital was just down the road a bit and had plenty of babies). She sounded just like that little old lady who says, "Where's the beef?" Everyone around us did his best to keep from howling out loud because at this point the performance was sort of on the solemn side, all serious and such. I myself was snorting into my elbow trying to be adult like but not really succeeding.

Love, Your girl who knows how to snort into her elbow

MARCH 1946

25 March 46

Dear Mother, Dad & kids,

Well things are about the same. I got two letters from you today also one from Mary Ann. I've been working pretty hard and the troop is coming along just fine.

I've been going into Linz once in a while at night. They have some swell officers' clubs in there with excellent music, American & Scotch whiskey, and champagne, cognac, etc. Since we are about the only troops here almost all the officers there are from our outfit.

The weather has been perfect and every day after retreat I go out and ride for an hour or so. We have 5 horses in our troop.

Those applications for Regular Army that I sent here never arrived so they'll have to use the ones in Washington.

Don't bother to send any whiskey; my room looks like a liquor store. We get plenty of American whiskey—good stuff for about $2 a fifth.

Enclosed are some photos of where my men live, its really a huge place. I'm thinking of moving to another castle, which is right on the Danube in a resort town.

I'm sure sorry I missed Dave—I wrote him on the boat.

I saw in the paper that officers and enlisted men are going to wear the same uniform—I guess that's okay, I'll save money anyway.

I saw Col. Fitzpatrick in the club in Linz last night—he hasn't changed a bit. He is the big Irishman I told you about.

Its only 7 o'clock but I was out late with Brooks last nite and have to get up at 5:30 so I'm going to bed.

Love to all, Jud

PS-New APO, 174, please notify papers & magazines.

December 20, 2014

Dear Dad,

It seems from the time between letters and your reference as to missing Dave that you finally got to go home. There are also a number of souvenir photos at dance clubs from Tulsa and one from New York at the Café Zanzibar on Broadway at 49th Street. You look happy and relaxed, quite full of yourself.

I finally graduated. We had a great time with nearly all the family there. As I was waiting to walk down the aisle and accept my long-awaited college diploma, I thought about all the graduations in the past. All of your grandchildren are college graduates, Dad. There were so many days I was filled with relief that my kids did not stray. Parenting is hard; parenting is exhausting, yet parenting makes you a better person. We partied that night at the Clubhouse complete with our own bowling alley, plenty of food, and an arcade. The next morning we had breakfast at Cracker Barrel, one of Mother's favorite places. Doug was such a good son, he used to take Mother there every Sunday for breakfast. She nearly always got the "Wild Maine Blueberry Pancakes" and undoubtedly she would share with the server, "I'm from Maine." She and I liked to go there and split the Baked Apple Dumplin. It was good for the soul for all of us to have "breakfast with Mother" one more time.

Afterward we headed to Savannah to spend the day and one night. We took a horse and buggy tour and ate dinner at the Pirates' House.

Your great grandson Travis as a pirate! Talk about full of it!!! Must be genetic!!

I really missed not having you and Mother at my graduation weekend. For some reason I feel flat now.

Love, Your girl without any parents

December 25, 2014

Merry Christmas Dad,
    We miss you.

Love, Your girl who has three wonderful grandchildren, Skylar, Travis and Ross

GENERAL JUDSON F. MILLER & KATHY WILLIAMS

Café Zanzibar

Great grandson Travis the pirate!

Merry Christmas from Skylar, Travis and Ross

APRIL 1946

10 April 1946

Dear Mother, Dad & kids,

Haven't had any mail for quite a while, I don't know why it's so slow.

Last weekend Ziggenhem and I really had a swell time. We went up into the mountains to a ski resort. It is really a place. It's a big hotel on top of a mountain 5200 ft. high. The ride up to the hotel on the cable car was quite a thrill. The place is civilian and we really got away from the army, it was wonderful. We took my valet, Otto, along with us. He is another Pancho. He was in the German ski troops and wounded 7 times in Russia. Before the war he was a ski instructor at a resort in Bavaria. We got along pretty good on our skiing for beginners. It's so warm up there you wear a bathing suit or shorts to ski in. Enclosed is a medal from the place, every resort has them and the idea is to get as many of them as possible to show where you've been and make a string of them which is hung from your belt. I got a good sunburn because the sun is so bright. It's really a wonderful sport. We're going again this weekend and improve our skiing. We get there Sat. afternoon, go to the dance that night, and ski Sunday. I'll try to take some pictures this time. The people up there really enjoy themselves, they are all very friendly. I never saw so many beautiful girls.

Tomorrow I'm going to the other side and talk to a Russian officer about our border control. I don't know what kind of a reception I'll get.

Enclosed is a clipping about our outfit from the Calvary Journal to which I subscribe. I also take the Army Navy Journal. Both are very good military magazines.

I read in the Stars and Stripes where they are going to start reducing officers down to and including captains but Purple Heart holders would get more consideration.

We're allowed to bring cars over now, so I wish you would do something for me. Write me the prices of the convertible coups of the following makes and how long I'd have to wait to buy each one—Ford, Mercury, Buick, Chrysler and Cadillac. Also about some packages with smokes and white underwear and a lastex (if available) bathing suit, I want to swim and I don't have a suit—size 34 is okay. Please write often.

Love, Jud

January 16, 2014

Dear Dad,

I can't believe this was your last letter in the box. It seemed like as long as I kept answering your letters, you would continue to write to me. Therefore I am having a hard time writing this letter. I do not want our written communication to end; it has given me great comfort to write to you, to dig into the boxes and learn about your life before I was conceived; the time you became a man. You loved cars as much as I love cats and you were an incredibly brave 20 year-old.

I want to be able to push aside the memories of that night you called me from Tacoma thinking you were back in World War II. "Kathy, is that you? Thank God. I am trapped in Company Headquarters and I can't get out." Now that I have read all your letters, I can understand the urgency and sense of craze in your voice. You survived three international wars but you could not escape the memories. I tried to reassure you it was going to be OK but the reality was I had lost Mother recently and I could see that I was losing you to a kind of Madness. I felt Mad myself at that moment.

Shelley was the first to reach you, then I arrived. It appeared the oxygen you were supposed to be getting was insufficient but I believe there was more going on. Were you wrangling with past demons? Perhaps recalling the 11 year-old boy you were forced to shoot, the unforgettable sound of the explosions, or the hunger pains as you lay in your tank trying to sleep. For a brief respite you were having fun again chasing Nazi women. Were you re-living the time spent drinking champagne on the Rhine?

But after 85 years of life your death was inevitable.

We buried your ashes next to Juddy's, honored your wishes for no obituary and no funeral. Instead that day we cooked your spaghetti, drank wine, and tried to fill the holes in our hearts.

Love, Your girl with a permanent hole in her heart

GENERAL JUDSON F. MILLER & KATHY WILLIAMS

Our last picture the same week you went into the hospital

EPILOGUE

May 31, 2015

Dear Dad,

When I asked you how you got your Silver Star in Vietnam you replied "you never leave anyone behind." You were referencing a phrase in the Soldier's Creed; never leave a fallen comrade. You not only believed in the Soldier's Creed, you lived it. And despite the fact that you have left the physical world, your presence continues to be a powerful and commanding force in my life. It is as if you have strategically placed certain people and events in my path that gave me direction and perseverance. The book is finished and I have been occupied with trying to find an agent or a publisher. Once again something is going on that is weird. No, it does not have anything to do with a cat. Remember that I do not buy into hocus pocus, "Luke it is your destiny," world religions, and the like. That does not mean that I am closed minded to the possibilities of the great unknowns, but I am pretty skeptical. I like logic and reality.

In my quest for pursuing publication I met John Edwards through the Blue Ridge Writers Group, my new local writing group. John recommended an editor named Alice Eachus who in turn recommended Bob Babcock with Deeds Publishing. There were 2.7 million Americans who served in Vietnam and today it is estimated that only 850,000 veterans are still alive. Bob Babcock served as a Second Lieutenant for the Second Brigade in Vietnam in 1966 and you were his Brigade Commander! Bob receives an average of 3 query letters a week and I received around 30 polite rejections but eventually we found each other. Can you believe it, Dad? Here was his reply:

Kathy,

Thank you for your inquiry. Military memoirs are among my favorite types of books to publish. As a Vietnam vet myself, and a WWII historian, I always love to read personal accounts from WWII vets. If you will forward me your manuscript in a Word document, I will be happy to review it and then we can discuss it.

Wait a minute! I may know your dad! I served under COL Jud Miller as our brigade commander in Vietnam - 2nd Brigade of the 4th Infantry Division in 1966-1967. I stayed in touch with him for many years after we came home and saw him in the Fort Lewis area around 1996 as I recall.

When would be a good time for us to talk? I now live in Athens, GA, having moved here recently from East Cobb County, north of Atlanta.

Deeds not Words!
Bob Babcock
CEO Deeds Publishing LLC

So Dad, I would really like to know just what is going on in the great beyond. Not only did he know you, he kept in contact with you after Vietnam.

Kathy,

It is a small world today. I just received this picture from Bill Saling, taken in 2013. From left to right are Jim Stapleton, Bill Saling, and Bob Babcock - we all served in 1-22 Infantry of 4ID in Vietnam in 1966-1967 with your dad as our brigade commander. All three of us live in Georgia - Jim in Atlanta, Bill in Big Canoe, and me in Athens. I responded to Bill and Jim, telling them about us just talking.

Bob Babcock

As I was going through the boxes one last time I came across two emails, the only copy of emails in the boxes. And guess who wrote one of them? Bob Babcock. Evidently he was with you when you went to visit the Vietnam Veterans Wall in 1994. The email from Bob was sent on November 1998, *17 years before* all of this was made possible. What led you to copy, save and put his email in the boxes? Some questions will always be left unanswered but I can assure you that you have not been left behind. Bob is carrying you now in his heart and I know he is the right person to help us tell our story.

To: Judd@seanet.com
Date: Tuesday, November 03, 1998 8:19 PM
Subject: 2nd Brigade, 4th Infantry Division – need some input

Jud, I have been invited to be the guest speaker at the 2nd Brigade Ball at Fort Hood, TX on 23 November 1998. I am eagerly anticipating the opportunity to revisit the brigade that I served with from Thanksgiving week of 1965 until I returned from Vietnam in July 1967.

They have asked me to talk about 'Army Values' that were applicable when I was in the Army that are still applicable now. I have some thoughts but would appreciate it if you, as a former 2nd Brigade commander (and the best one I ever knew about) had any words of wisdom that I could pass on to the current officers and NCOs of the 2nd Brigade.

Besides words of wisdom on values, I am also interested in human interest stories that you can recall. I will most likely use the story we have discussed about the naming of Camp Enari and the two Collins men who were killed before a name was decided on.

If you have anything that I could pass on to the current soldiers, I would be most grateful. Hearing from a LT who happens to be president of the National 4th Infantry Division Association is nice but they would also like to hear words of wisdom and anecdotes from the man who led the brigade into combat for the first time (and I guess the only time the brigade has fought a war).

Hope all is well with you. I am staying very busy. Will be going to Washington, DC to the Wall on Veterans Day. I often think of the day that you, Rawhide, and I went to the Wall together – and also think of our day together at the Holocaust Museum. Looking forward to seeing you at Fort Hood next July at our 4IDA reunion.

Bob Babcock "Deeds Not Words"
Industry Segment Executive
IBM North America Small & Medium Business
Atlanta, GA
E-mail: babcock@us.ibm.com

~

To:Bob Babcock babcock@us.ibm.com
Date:Monday, November 09, 1998 2:25 PM
Subject:Re: 2nd Brigade, 4th Infantry Division – need some input

Hi Bob. Thanks very much for your message. I know that you will have a great time at Fort Hood.

I really don't know that I have any great words of wisdom for you except to highlight the absolute necessity for INTEGRITY in development of the chain of command. Unless commanders at all levels can rely completely on information and reports they receive from their subordinates, there is no way that they can function effectively as leaders. Accordingly, each leader must be completely honest and forthcoming in the information he provides to his superior – no inflated body counts, no false readiness reports, no concealment of unfavorable information and soforth. The commander must know that the information on which he bases his decision is accurate and complete and not false representations by less than honest self-serving subordinates. During my tenure in command of the 2d Brigade in training at Fort Lewis and in combat I always felt that I was able to rely on the integrity of my subordinates in providing me with complete and accurate information.

I am sure that you recall that just before the brigade shipped out, virtually all 5,000 men in the Brigade Task Force were given leave. Just after everyone went on leave, there was a nationwide airline strike and some predicted that hundreds of soldiers would miss shipment to Vietnam. To the contrary, all but 5 or 10 soldiers made it back on time by hitch hiking, taking the bus, using Canadian airlines or other means to overcome the lack of air transportation in the US.

I hope that the forgoing is of some help to you.

Please give my best regards to the 2d Brigade. My command of this great outfit was the highlight of my 34 years in the United States Army.

I'll see you in July, 1999 at the reunion.

Jud

The weird is not over yet. Bob related to me the story behind the naming of your camp in Vietnam.

Major General Collins, the Division Commander, instructed you to name the camp after the first GI killed by hostile fire in order to honor the soldier. The first GI killed was PFC Albert Collins. You were sensitive that General Collins would not want any false perceptions that the camp was being named after the General instead of PFC Collins and proposed naming the camp after the first officer killed in action instead. This was agreed upon. However, the first officer KIA ended up being Lieutenant Richard Collins. Therefore the plan changed to naming the camp after the first posthumous recipient of the Silver Star, no matter the rank or name. This turned out to be Lieutenant Mark Enari. Thus your camp became known as Camp Enari.

This past Memorial Day, May 25, 2015 Bob Babcock attended a service in Marietta, Georgia. The general speaking that day decided to tell the story behind the naming of Camp Enari because the first GI killed, PFC Collins, is buried at the National Cemetery in Marietta! Bob sent me the following picture and caption in front of PFC Collins' grave.

Jim Stapleton, Bob Babcock, Mike Hamer - all served in Vietnam at same time as your dad. Jim and I were in the same battalion in his brigade, Mike was in another brigade.

Well Dad, what do you think about all of this? Some will say it is a sign from God, others just coincidence. Maybe you are still giving orders from the great beyond. I can see that happening. Who knows? I guess what is really important is that you live on, if nowhere else but in the shared memories of people who never knew each other until now.

Love, Kathy "Martha Gasoline"

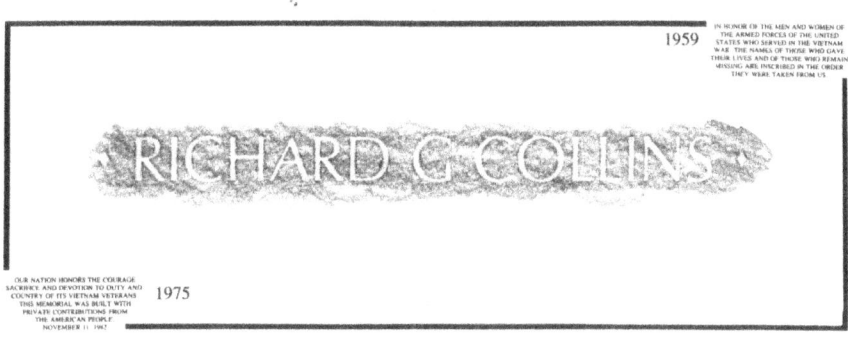

GENERAL JUDSON F. MILLER & KATHY WILLIAMS

1959

IN HONOR OF THE MEN AND WOMEN OF
THE ARMED FORCES OF THE UNITED
STATES WHO SERVED IN THE VIETNAM
WAR. THE NAMES OF THOSE WHO GAVE
THEIR LIVES AND OF THOSE WHO REMAIN
MISSING ARE INSCRIBED IN THE ORDER
THEY WERE TAKEN FROM US.

MARK N ENARI

OUR NATION HONORS THE COURAGE
SACRIFICE AND DEVOTION TO DUTY AND
COUNTRY OF ITS VIETNAM VETERANS
THIS MEMORIAL WAS BUILT WITH
PRIVATE CONTRIBUTIONS FROM
THE AMERICAN PEOPLE
NOVEMBER 11, 1982

1975

AFTERWORD

My parents met in Austria a year after the war. She was his nurse when he was hospitalized for a brief illness. They were married in Austria in April 1947, she was 23 years old and he was just 21 years old. I was their first child, born in 1949.

My mother's wedding gown was made of parachute nylon. They honeymooned in Paris, Bavaria, and the French Riviera. My Dad went on to serve in Korea and Vietnam. He retired as a General, then received a law degree and practiced law in Tacoma, Washington.

Wedding Day Austria 1947

First family photo with oldest daughter Kathy

My elegant mother who believed I could write

PHOTOGRAPHS

Jump school Ft Bragg, NC

A car at long last!!!

In the mountains of Korea

Korean devastation

The bleakness of the Korean War

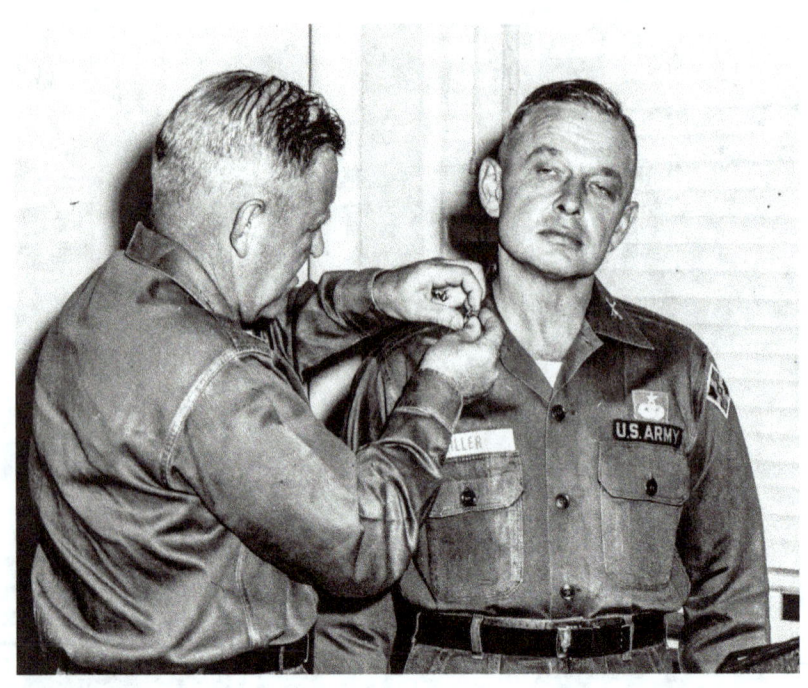

6357-251-01 of 2/AM 66   5 May 66
4th Inf Div   Ft Lewis, Wash

General Creighton W. Abrams Jr. Army Vice Chief of Staff pins the Colonel's insignia on Lt Col Judson F Miller, the 2nd Brigade commander at a surprise promotion ceremony held at Brigade Headquarters at Ft Lewis, Washington.

U.S. ARMY PHOTOGRAPH: PHOTOGRAPHER
SP/5 GEORGE R CASH
124TH SIG BN CO C

THE COMMANDING GENERAL AND SPECIAL STAFF

JUNE 1967

First Row: LTC Robinson AG, LTC Crizer G2, LTC Morley G1, BG Walker ADC-A, MG Peers CG, BG Ryder ADC-B, COL Miller CS, LTC Lay G3, LTC Richards G4, LTC Allyn G5

Second Row: LTC Friendly IG, LTC Hett CHAP, LTC Schug SJA, LTC Cullen PMO, LTC Delbridge ENG, LTC Holloman AVN, LTC Spitz SIG, LTC Lennon ASST FSC, LTC Hillis ALO

Third Row: LTC Peard SURG, LTC Henderson CML, MAJ Smith ASST AVN O, MAJ FOSTER PSYOPS, MAJ Zenk IO, MAJ Neal ASST CS.

U.S. ARMY PHOTOGRAPH
4TH INF DIV INFORMATION OFFICE
PHOTOGRAPHER: SP4 RONALD SATO

Disembarking in Vietnam

Welcome from General Westmoreland

CHANGING HANDS — Colonel Judson F. Miller, left, passes the brigade colors and with them the command of the 4th Division's 2nd Brigade to Colonel James B. Adamson in ceremonies Sunday afternoon. Colonel Miller is now serving as division chief of staff. (USA Photo by SSgt. William J. Whitis)

COLONEL MILLER
New Chief Of Staff Receives Silver Star

Plei Djereng — Colonel Judson F. Miller, newly named 4th Division chief of staff and former commander of the 2nd Brigade, has received the nation's third highest award for heroism.

Major General William R. Peers, 4th Division commander, presented the Silver Star to Colonel Miller during ceremonies Sunday at the 2nd Brigade forward base camp.

Colonel Miller, a native of Oklahoma, was cited for his actions in leading his brigade against a hostile force November 10-13. Without regard for his own safety, the colonel he could establish communications with his commanders.

The following day, he led

(Cont'd on p-6, Col. 1)

MEDAL FOR COLONEL — Major General William R. Peers, commander of the 4th Division, pins the Silver Star on Colonel Judson F. Miller during ceremonies Sunday at the 2nd Brigade's forward command post heliport. (USA Photo by SSgt. William J. Whitis)

Colonel Miller Receives Silver Star

(Cont'd from p-1, Col. 2) the 1st Battalion, 12th Infantry, in heavy fighting until a secure perimeter was established. During this time he continually exposed himself to fire which inspired his commanders and men.

Before assuming command of the 2nd Brigade last February, he served with Headquarters, U.S. Strike Command, Florida. Other assignments have been with 4th Cavalry Group, Europe; 82nd Airborne Division; 187th Airborne Regimental Combat Team, Korea; Army Armor School, Ft. Knox, Ky.; and 14th Cavalry Regiment in Germany.

The new chief of staff, who wears senior parachutist wings, has attended the Command and General Staff College and the Army War College. Colonel Miller takes pride in the fact that he acquired a bachelor of arts degree and masters degree— all on off duty hours while in the Army.

His previous decorations include the Bronze Star for Valor and Oak Leaf Cluster, Joint Service Commendation Medal, European Theater Ribbon with five battle stars, Korean Campaign Medal with six battle stars, and the Purple Heart.

Troop review with General Westmoreland

Tulsan Col. Judson Miller Gets Call: 'I'm Not Dead'

PLEIKU, Viet Nam (AP)—Lt. Col. James R. Lay telephoned Col Judson F. Miller Thursday to deny a report that he was dead.

Lay, of Westminster, S.C., is commander of the 1st Battalion, 12th Infantry, 2nd Brigade, 4th Infantry Division. Miller of Tulsa, is the brigade commander.

(Miller, son of Mrs. H. F. Miller of 3240 S. Zuni Place, is a 23-year Army veteran. He fought in Europe during World War II and is also a Korean War veteran. He has been in Viet Nam since last summer.)

Lay, in the field with his battalion, told Miller he had used his transistor radio to listen to an English news broadcast over Radio Peking.

He said the Red Chinese announcer reported that in heavy fighting West of Pleiku one complete battalion of the 4th Division had been wiped out with more than 1,000 Americans killed, including the battalion commander

"I told Lay I was very sorry about that," Miller said.

Lay replied, "Well, don't expect too much from us in the future."

Lay's battalion suffered light casualties while repulsing repeated attacks by North Vietnamese soldiers Saturday night. So far, 88 enemy bodies have been found in the foliage surrounding the battalion perimeter.

Dale Robertson, movie actor and fellow OMA cadet, visits Vietnam

25 Nov 66

Dear Dave,

Thanks for your letter and copies of the clipping.

We continue to be where the action is. We get hard contact with the NVA almost every day and are racking up a very nice count of Charlies. I have been given priority (by Westy) on use of the B52s to hit areas where we believe Charlie to be located before we move in. This has worked well as they really tear up their bunkers and make life much easier for the infantry.

In addition to my own units I have several CIDG units (Vietnamese led by Special Forces) attached. I don't think too much of the Green Berets. They manage to get themselves in all sorts of trouble and then we have to bail them out.

I have been notified by the division commander that I will become Division Chief of Staff on

- 2 -

January 1. I dislike leaving the brigade but from a career standpoint it is number 1. This job normally goes to the senior colonel in the division — I am the most junior. Gen Collins really seems happy with my work so if I don't lose a "battalion" in the next month I should be OK.

Have heard that I am to receive a decoration from Premier Ky Monday or Tuesday — Hot Dog!

As you can see by the enclosed I am a genuine helicopter hero. The first and second awards are presently pending — all that is required for this is a certain number of missions.

When my acts of heroism appear on page 1 the enclosed photo may be better than what they have. This is my only copy.

-3-

Tell Mother the package arrived in good shape and I appreciate it very much.

Since we became involved in this heavy action we have had swarms of press & TV types here. Most of them are pretty straight but several are always trying to get me to say (on film) that the NVA is in Cambodia — of course we all know that this ain't so. Right now I have Peter Arnett, Morey Sapper and Horst Faas plus some others here. I have used up a year's liquor ration entertaining these birds. My favorite correspondent is Michele Ray who spent a week with us. I appear in the movie she is making.

Write
Jud.

Enclosed are some negatives Mother may want prints of

ABOUT THE AUTHOR

Kathy was born in Ft Bragg, NC in 1949. Her parents met and married at the end of WWII in Austria. Her father was an Army Captain, her mother an Army nurse. A childhood spent as a traveling "Army Brat" enriched her life and widened her horizons.

Her early careers included an unsuccessful stint as a jockey followed by a little more productive race horse training business. She then spent nearly 22 years in the convenience store business. Inspired by her children and parents, she completed her BA in International Studies with a minor in Spanish at Georgia Southern University after fourteen years of night classes. During her last year at Georgia Southern she wrote her first book, *Dear Dad*.

Writing has always been a part of her life that she wanted to pursue. She was the editor of her high school paper. Upon her retirement from Flash Foods in 2014, she relocated to Blue Ridge, GA to be close to her children and grandchildren. Here she has engaged in many volunteer opportunities, including becoming the newsletter editor for the Benton MacKaye Trail Association. She also participates as a volunteer at the Blue Ridge Community Theater and is a Friend of the Fannin County Library as well as a member of the Blue Ridge Poets and Writers Group.

Some of her favorite things include running, triathlons, hiking, anything *Star Wars* or *Lord of the Rings*, Stephen Colbert and Bill O'Reilly, animals, Jeopardy, history, travel, reading and the Atlanta Falcons.

www.ingramcontent.com/pod-product-compliance
Lightning Source LLC
Chambersburg PA
CBHW071725080526
44588CB00013B/1897